Disclaimer

The publisher of this book is by no way associated with the National Institute of Standards and Technology (NIST). The NIST did not publish this book. It was published by 50 page publications under the public domain license.

50 Page Publications.

Book Title: Smoke Plume Trajectory from In Situ Burning of Crude Oil in Alaska - Updated Simulation Results

Book Author: William D. Walton; Kevin B. McGrattan;

Book Abstract: The purpose of this report is to update calculations, originally performed in 1993, that predict the downwind extent of smoke particulate from hypothetical in situ burns of spilled crude oil in Alaska. The reason for the update is that the National Ambient Air Quality Standards (NAAQS) have changed since 1993. These standards formed the basis for establishing ‚safe distancesŠ for separating potential burning sites from populated areas in Alaska.

Citation: NIST TN - 1706

Keywords: ALOFT, fire modeling, in situ burning, oil fires

NIST Technical Note 1706

Smoke Plume Trajectory from In Situ Burning of Crude Oil in Alaska – Updated Simulation Results

William Walton
Kevin McGrattan

NIST
National Institute of
Standards and Technology
U.S. Department of Commerce

NIST Technical Note 1706

Smoke Plume Trajectory from In Situ Burning of Crude Oil in Alaska – Updated Simulation Results

William Walton
Kevin McGrattan

Fire Research Division
Engineering Laboratory

July 2011

U.S. Department of Commerce
Gary Locke, Secretary

National Institute of Standards and Technology
Patrick D. Gallagher, Director

National Institute of Standards and Technology Technical Note 1706
Natl. Inst. Stand. Technol. Tech. Note 1706, 51 Pages (July 2011)
CODEN: NTNOEF

Abstract

The purpose of this report is to update calculations, originally performed in 1993, that predict the ground level concentration of smoke particulate from hypothetical *in situ* burns of crude oil spills in Alaska. The reason for the update is that the National Ambient Air Quality Standards (NAAQS) have changed since 1993. These standards formed the basis for establishing "safe distances" for separating potential burning sites from populated areas in Alaska. A total of 28 scenarios were examined, with 2 types of crude oil, 3 wind speeds, 3 fire areas, and winter and summer meteorological conditions. The downwind extents of 3 different concentrations of fine smoke particulate ($PM_{2.5}$) are presented. The concentration thresholds represent different versions of the NAAQS 24-hour average particulate standard.

Executive Summary

The purpose of this report is to update calculations, originally performed in 1993, that predict the ground level concentration of smoke particulate from hypothetical *in situ* burns of crude oil spills in Alaska. To establish "safe distance" criteria for populated areas, the Alaska Department of Environmental Conservation (ADEC) adopted the existing 24-hour National Ambient Air Quality Standards (NAAQS) for PM_{10}. The primary reason for the updated calculations is that the NAAQS have changed since 1993. In 1993, the 24-hour standard for PM_{10} was 150 $\mu g/m^3$. Several years later, in 1997, an additional standard was added for $PM_{2.5}$. The original 24-hour standard for $PM_{2.5}$ was 65 $\mu g/m^3$, which was changed in 2006 to its current value of 35 $\mu g/m^3$. It is possible that the 24-hour standard for $PM_{2.5}$ could decrease yet again to 25 $\mu g/m^3$.

ALOFT-FT (A Large Outdoor Fire plume Trajectory, Flat Terrain) is the smoke dispersion model, developed at the National Institute of Standards and Technology. The model predicts the 1-hour average ground level concentration of smoke particulate from large outdoor fires. The model uses a 1-hour rather than a 24-hour average because the duration of *in situ* burns is expected to be nominally a few hours, and the model cannot accommodate variations in meteorological conditions that would be expected over a 24 hour period. Using the most recent version of the ALOFT-FT model, the extents of 1-hour averaged, ground level $PM_{2.5}$ concentrations of 25 $\mu g/m^3$ and 35 $\mu g/m^3$ are predicted to be less than 10 km (6 mi) downwind of the fire for all scenarios considered. The extent of 150 $\mu g/m^3$ (PM_{10}) predicted in 1993 was 5 km (3 mi).

Table of Contents

Abstract i

Executive Summary iii

Table of Contents v

List of Figures vii

List of Tables vii

List of Abbreviations ix

Acknowledgments xi

Introduction 1

Air Quality Standards Relevant to *In Situ* Burning 2

Important Operational Considerations for *In Situ* Burning 4

    Burn Area 4

    Burning Rate and Soot Yield 6

ALOFT-FT Model 8

    Mathematical Model 8

    Validation 9

    Model Inputs 10

    ALOFT-FT Version History 10

    ALOFT-CT (Complex Terrain) 11

ALOFT-FT Modeling Results 13

Discussion of ALOFT-FT Results 16

Conclusion 19

References 20

Appendix 23

List of Figures

Figure 1. The basic catenary shape of a tow boom. 6
Figure 2. Comparison of ALOFT-FT predictions (shaded contours) with field measurements (numbers) for the ACS Burning of Emulsions Experiment. Units are $\mu g/m^3$. 18
Figure 3. Temperature profiles for the four meteorological conditions (McGrattan *et al.*, 1993).23

List of Tables

Table 1. Properties of oils used in the simulations (McGrattan *et al.*, 1993). 7
Table 2. Summary of fire scenarios. 14
Table 3. Predicted extents of three different ground level hour-average concentrations of smoke particulate from a variety of fires under different weather conditions. 15

List of Abbreviations

ADEC	Alaska Department of Environmental Conservation
ALOFT-FT	A Large Outdoor Fire plume Trajectory – Flat Terrain
ALOFT-CT	A Large Outdoor Fire plume Trajectory – Complex Terrain
ARRT	Alaska Regional Response Team
bbl	barrels
CAA	Clean Air Act
CFR	Code of Federal Regulations
CRADA	Cooperative Research and Development Agreement
EPA	Environmental Protection Agency
FR	Federal Register
ISB	*In Situ* Burning
$\mu g/m^3$	micrograms per cubic meter
LES	Large Eddy Simulation
NAAQS	National Ambient Air Quality Standards
NIST	National Institute of Standards and Technology
NOBE	Newfoundland Offshore Burn Experiment
NRT	National Response Team
RAM	Real-Time Aerosol Monitor
RRT	Regional Response Team
$PM_{2.5}$	Fine Particulate Matter, 2.5 micrometers (microns) or less in size
PM_{10}	Coarse Particulate Matter, 10 micrometers (microns) or less in size
SIP	State Implementation Plan
S&T	Science and Technology

Acknowledgments

This study has been performed as part of a Memorandum of Agreement between the Alaska Department of Environmental Conservation (ADEC) and the National Institute of Standards and Technology (NIST). Technical assistance has been provided by ADEC staff members, in particular Larry Dietrick, Alice Edwards, Gary Folley, Lawrence Iwamoto, and Barbara Trost.

Introduction

As part of their effort to assess the impact of smoke plumes from *in situ* burning (ISB) on nearby populations, the Alaska Regional Response Team (ARRT) and the Alaska Department of Environmental Conservation (ADEC) established a Cooperative Research and Development Agreement (CRADA) with the National Institute of Standards and Technology (NIST) in 1993. The Fire Research Division at NIST conducted laboratory-scale and large-scale fire experiments, measuring the heat release and smoke production rates of various types of crude oil. In addition, NIST developed a numerical model, A Large Outdoor Fire plume Trajectory (ALOFT), to predict the downwind, ground level concentration of smoke particulate from a large oil fire. The results of this work were documented in a report that was delivered to ADEC (McGrattan *et al.*, 1993). Based on the model predictions, ADEC developed an initial set of ISB Guidelines that were approved by the ARRT in May 1994. At that time, the 24-hour National Ambient Air Quality Standard (NAAQS) for PM_{10} was 150 $\mu g/m^3$. A panel of experts determined that the 24-hour average should be used to evaluate the emissions from an *in situ* burn as opposed to the annual average value.

Later in the decade, in 1997, an additional standard for fine particulate ($PM_{2.5}$) was added to the NAAQS. The original 24-hour standard for $PM_{2.5}$ was 65 $\mu g/m^3$. The ALOFT-FT model -- having been validated using data from actual controlled burns off the coast of Newfoundland, Canada, on the North Slope of Alaska, and in Mobile Bay, Alabama -- was used to recalculate safe distances for *in situ* burning of oil based on the new $PM_{2.5}$ standard. Revised guidelines were developed to address the change in the NAAQS and the revision was forwarded to the ARRT for review and approval. Before the ARRT review was completed, the 24-hour standard for $PM_{2.5}$ changed in 2006 from $65\mu g/m^3$ to $35\mu g/m^3$. Several organizations within the State of Alaska questioned the validity of the original safe distance calculations that were based on the 1993 NIST study. These distances remained the same despite the change in the NAAQS.

The purpose of this report is to repeat the ALOFT-FT calculations reported in the 1993 NIST study, only now with the results reported in terms of the new 24-hour standard for $PM_{2.5}$, $35\mu g/m^3$. In addition, the fire sizes and other assumptions made for the 1993 study are re-assessed in light of the roughly 400 *in situ* burns conducted during the recovery operations for the *Deepwater Horizon* spill in the Gulf of Mexico in 2010 (Allen, 2010).

Air Quality Standards Relevant to *In Situ* Burning

The Clean Air Act (CAA) is the primary Federal air quality law and includes the National Ambient Air Quality Standard (NAAQS). The CAA is implemented by states and localities through State Implementation Plans (SIPs). The Environmental Protection Agency (EPA) is responsible for setting and periodically reviewing the NAAQS for certain pollutants including particulate matter. The term particulate matter (PM) includes both solid particles and liquid droplets found in air. Many man-made and natural sources emit PM directly or emit other pollutants that react in the atmosphere to form PM. These solid and liquid particles come in a wide range of sizes.

The first national air quality particulate standards[1] were established in 1971 and were revised in 1987 to regulate particles with an effective aerodynamic diameter smaller than or equal to 10 micrometers, PM_{10}. The standard was intended to regulate inhalable coarse particles but included both course and fine particles with a 24-hour average of 150 $\mu g/m^3$. Particles less than 10 micrometers in diameter (PM_{10}) pose a health concern because they can be inhaled into and accumulate in the respiratory system. Particles less than 2.5 micrometers in diameter ($PM_{2.5}$) are referred to as "fine" particles and are believed to pose the largest health risk. Because of their small size (less than one-seventh the average width of a human hair), fine particles can lodge deeply into the lungs. In 1997, the EPA established annual and 24-hour standards for $PM_{2.5}$ for the first time. The 24-hour standard was 65 $\mu g/m^3$. In 2006, the EPA reduced the 24-hour NAAQS for $PM_{2.5}$ to 35 $\mu g/m^3$.

Health studies have shown a significant association between exposure to fine particles and premature mortality (Pope *et al.*, 2002, 2006). Other important effects include aggravation of respiratory and cardiovascular illness, lung disease, decreased lung function, asthma attacks, and certain cardiovascular problems such as heart attacks and cardiac arrhythmia. Individuals particularly sensitive to fine particle exposure include older adults, people with heart and lung disease, and children. Recent community studies find that adverse public health effects are associated with exposure to particles at levels well below the previous PM standards for both short-term (from less than 1 day to up to 5 days) and long-term (from generally a year to several years) periods.

The 2006 NAAQS established a new form for the annual $PM_{2.5}$ standard. Areas will be in compliance with the new annual $PM_{2.5}$ standard when the 3-year average of the annual arithmetic mean $PM_{2.5}$ concentrations, from single or multiple community-oriented monitors, is less than or equal to 15 $\mu g/m^3$. The use of averages from single or multiple community-oriented sites is more closely linked to the underlying health effects information, which relates area wide health statistics to averaged measurements of area wide air quality. EPA believes this more protective annual standard, with the supplemental protection afforded by the 24-hour standard, which is directed at peak concentrations and localized hot spots, will provide a protective target that will reduce area-wide population exposure to fine particles.

---

[1] National Primary and Secondary Ambient Air Quality Standards for Particulate Matter, 52 FR 24663, July 1, 1987, as amended at 62 FR 38711, July 18 1997; 65 FR 80779, Dec. 22, 2000; 71 FR 61224, Oct. 17, 2006.

For the new 24-hour $PM_{2.5}$ standard, the form is based on the 98th percentile of 24-hour $PM_{2.5}$ concentrations in a year (averaged over 3 years), at the population-oriented monitoring site with the highest measured values in an area. The 24-hour standard will limit peak concentration in areas with high seasonal concentrations and in areas with localized hot spots due to particular sources. This form will reduce the impact of a single high exposure event that may be due to unusual meteorological conditions, and thus would provide a more stable basis for effective control programs. The percentile form compensates for missing data and less-than-everyday monitoring, thereby reducing or eliminating the need for complex procedures previously required for the $PM_{2.5}$ attainment test.

The 2006 rule revoked the annual (arithmetic mean) for PM_{10} due to a lack of evidence linking health problems to long term exposure to coarse particulate matter. The rule stating that the 150 $\mu g/m^3$ 24-hour average is not to be exceeded more than once per year on average over three years remained.

It should be noted that according to 40 CFR Part 50, Appendix N, "Data from exceptional events, for example structural fires or high winds, may be given special consideration. In some cases, it may be appropriate to exclude these data in whole or in part because they could result in inappropriate values to compare with the levels of the $PM_{2.5}$ NAAQS."

In 1995, the National Response Team (NRT) Science and Technology (S&T) Committee published a guide on the *Applicability of Clean Air Ambient Air Quality Regulations to the In Situ Burning of Oil Spills*. The NRT is an organization consisting of participants from 15 Federal departments and agencies responsible for coordinating emergency preparedness and response to oil and hazardous substance pollution incidents. A group of experts empanelled by the NRT S&T committee examined the toxicity of the pollutants in the NAAQS that would apply to *in situ* burning. They determined that soot posed the greatest potential threat to the public. Further they chose the 24-hour average as the standard to use. For this reason, the modeling has focused on predicting soot particulate concentration downwind of a burn. However, assuming the yield of another potential pollutant is known, the model results can be reformulated accordingly. It is assumed that all combustion products are transported together, at least in the vicinity of the burn. Over a time period of a few days, smoke particulate will settle out of the plume at a slow rate because of its very small effective aerodynamic diameter. The smoke plume dissipates to concentrations far below the NAAQS limits before any significant settling of particulate can occur (Evans et al., 2001).

For the purpose of evaluating air quality impacts from smoke plumes from *in situ* burning, modeled ground level particulate matter concentrations are compared to the 24-hour $PM_{2.5}$ NAAQS rather than the PM_{10}. The particulate matter size range from *in situ* burning of crude oil is determined by the oil type and the burn conditions. Measurements of the particulate size distributions from burns of two types of Alaskan crude oils indicate that the $PM_{2.5}$ yield is approximately 75 % of the PM_{10} yield. This means that 75 % of the PM_{10} mass is $PM_{2.5}$. As a result, the current NAAQS 24-hour average standard for $PM_{2.5}$ (35 $\mu g/m^3$) is more stringent than the corresponding value for PM_{10} (150 $\mu g/m^3$).

Important Operational Considerations for *In Situ* Burning

The phrase *in situ* burning, or ISB, refers to the burning of an oil spill in the location where it is spilled. The phrase *in situ* is Latin for "in position" or "in place." Oil could also be burned by mechanically recovering the oil and transporting it to a site such as an incinerator for burning. Oil spills are most easily burned when the oil is "fresh," meaning that the oil has not been exposed to the environment for a long period of time. Over time oil begins to evaporate and the lightest molecular fractions which are the easiest to ignite evaporate first. Further, oil in contact with water may begin to mix with the water or emulsify. As the water content of emulsified oil increases, the oil becomes more difficult to ignite and burn. For these reasons the decision to conduct an *in situ* burn needs to be made rapidly or the opportunity to ignite the oil may pass.

In situ burning may be either intentional or unintentional. Fires can often be the direct cause or a consequence of the spill. For example, the oil tanker *Mega Borg* caught fire off the coast of Texas in 1990 during lightering operations. Oil burned both on the ship and on the water. The *Deepwater Horizon* platform caught fire off the coast of Louisiana in 2010 resulting in unintentional oil burning on the structure and on the water.

Usually the phrase *in situ* burning refers to an intentional burn of spilled oil. The most notable examples are the exploratory burns performed following the *Exxon Valdez* spill in 1989, and the extensive controlled burns following the loss of the *Deepwater Horizon* in 2010. In both events, oil was collected in booms and intentionally burned.

The smoke from a large oil fire includes carbon dioxide, water vapor, smoke particulate, carbon monoxide, hydrocarbons, sulfur compounds, oxides of nitrogen, and other aerosols and gases. The particulate is of the greatest interest in assessing potential health effects from exposure to the smoke because it has been shown (McGrattan *et al.*, 1997) that the most likely combustion product to violate ambient air quality standards is the particulate. Of the two types of particulate that are addressed by the NAAQS, $PM_{2.5}$ is more likely to be found in excess of the NAAQS 24-hour limit. Approximately 10 % (by mass) of the burned oil is converted to smoke particulate, of which roughly 75 % by mass is $PM_{2.5}$. Besides carbon dioxide and water vapor, none of the other by-products is generated at such a high rate.

There are three principle factors that determine the quantity of smoke particulate produced by an *in situ* burn. These are the average fire area, the average oil burning rate, and the average soot yield. The fire area is literally the area of the burning oil, not the area of the spill itself. The burning rate is the rate at which oil mass is consumed by the fire, and the soot yield is the mass fraction of the oil that is converted to particulate matter. Both the burning rate and soot yields are functions of the oil type and the burning conditions. The production rate of particulate is simply the burning rate multiplied by the soot yield.

Burn Area

The area of an unconfined oil burn is difficult to determine, especially if the oil is spilled on water. The reason is that the layer of oil thins out towards the edge of the slick, and eventually the heat loss to the water exceeds the heat necessary to sustain the burning.

The containment of oil can be achieved in one of two ways. The first way is to exploit a pre-existing manmade or natural barrier. These include, for example, coves, jetties, or piers. The areas of these burns are difficult to predict for the same reason as with an unconfined burn. The second way of controlling the area of an *in situ* burn is the use of a fire resistant oil spill containment boom. This is the method most commonly considered by response personnel and the method most commonly modeled in preplan exercises.

During the *Deepwater Horizon* incident in the Gulf of Mexico in 2010, *in situ* burning was used extensively. It is estimated that 41,300 m³ (260,000 barrels) of oil was burned (Lehr *et al.*, 2010). Most of the oil was burned using a 152 m (500 ft) long boom with a target "gap" ratio (swath to boom length ratio) of 0.3 or an opening of 46 m (150 ft). It was estimated that a single boom could hold 80 m³ to 160 m³ (500 bbl to 1000 bbl) with the oil one-third of the distance from the apex to the open end of the boom. Although the areas of the individual *Deepwater Horizon* burns are not given by Lehr *et al.* (2010), it appears that the burn areas used in this analysis are similar to those used during the spill response.

A typical fire-resistant boom is approximately 152 m (500 ft) in length. Two tow boats drag the boom through the water forming a U or catenary shape (see Figure 1). The open end of the U is typically maintained at roughly 46 m (150 ft). Oil is collected at the closed end of the U and the area of the oil depends on the quantity of oil in the boom and the speed of the tow vessels. Ideally the boom is towed at a rate to maintain maximum burn efficiency. The area of the oil within the boom can be determined knowing the shape of the boom and the centerline distance filled with oil (Allen, 1999).

The equation for the catenary shape naturally formed by a 152 m (500 ft) boom towed by two boats spaced 46 m (150 ft) apart is:

$$d = a\left[\cosh\left(\frac{w}{a}\right) - 1\right]; \quad a \cong 7.6 \text{ m } (25 \text{ ft})$$

where d is the distance from the apex and w is the perpendicular distance from the centerline (see Figure 1). The parameter a is a scaling factor. The area within the boom as a function of the distance from the apex is given by:

$$A(d) = 2wd - 2a^2 \sinh\left(\frac{w}{a}\right)$$

$$w = a \ln\left(\frac{d+a}{a} + \sqrt{\left(\frac{d+a}{a}\right)^2 - 1}\right)$$

The total area of the boom is approximately 1930 m² (20,800 ft²). The area of the boom from the apex to one eighth of the distance to the opening is approximately 232 m² (2,500 ft²), to one quarter of the distance is 465 m² (5,000 ft²), and to one half the distance is 930 m² (10,000 ft²). These are the three areas analyzed in this report. They span the most likely areas reported for the *Deepwater Horizon* (Allen, 2010). It should be noted that multiple *in situ* burns may take place at the same time. This was the case on some days during the *Deepwater Horizon* incident. The total smoke being dispersed is the sum of the individual fires occurring at the same time.

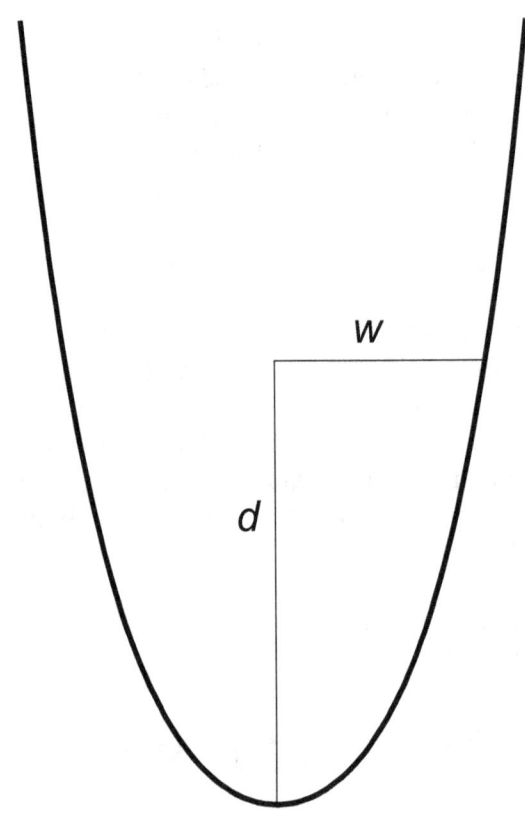

Figure 1. The basic catenary shape of a tow boom.

Burning Rate and Soot Yield

The primary combustion products from burning crude oil or any other hydrocarbon fuel are carbon dioxide and water vapor. Depending on the type of oil, a wide range of secondary combustion products are produced including carbon monoxide, sulfur compounds, carbon compounds, nitrogen oxides, unburned hydrocarbons, and soot. The soot or particulate matter in the smoke is primarily elemental carbon (Evans *et al.*, 2001).

Table 1 presents the burning rate, heat release rate and soot yield of particulate matter less than 10 μm and 2.5 μm in size (PM_{10} and $PM_{2.5}$) from two types of Alaskan crude oil. These measurements were first reported by McGrattan *et al.* (1993). The original ALOFT-FT calculations used these same burning rates. However, the 24-hour average NAAQS for PM_{10} in 1993 was 150 $\mu g/m^3$. In order to convert the yield for PM_{10} to a yield for $PM_{2.5}$, it is only a matter of multiplying the PM_{10} value for Cook Inlet crude by 0.70 and the value for North Slope crude by 0.75. These factors are based on the size distribution of the particulate measured by two types of 8 stage impactors.

Table 1. Properties of oils used in the simulations (McGrattan *et al.*, 1993).

Oil	Average Burning Rate $kg/(m^2 \cdot s)$ $[gal/(ft^2 \cdot h)]$	Average Heat Release Rate (kW/m^2)	Average Soot Yield PM_{10}	Average Soot Yield $PM_{2.5}$
Cook Inlet	0.056 ± 0.008 $[5.9 \pm 0.9]$	2180 ± 330	0.092 ± 0.006	0.064 ± 0.004
North Slope	0.051 ± 0.008 $[5.1 \pm 0.8]$	1960 ± 300	0.116 ± 0.010	0.086 ± 0.007

Note: the uncertainties are expressed as 95 % confidence intervals.

ALOFT-FT Model

ALOFT (A Large Outdoor Fire plume Trajectory, Flat Terrain) is a numerical model that predicts the downwind concentration of combustion products from large outdoor fires. ALOFT-FT is a public domain, personal computer implementation of the ALOFT model for flat terrain. This software was developed by the National Institute of Standards and Technology (NIST)[2].

Mathematical Model

The ALOFT-FT mathematical model assumes that the smoke plume consists of heated combustion products introduced into the atmosphere by a continuously burning fire and transported by a uniform wind field. A detailed description of the equations used in the ALOFT-FT model has been presented previously (McGrattan *et al.*, 1997). The model does not predict the behavior of the fire itself, but rather the plume of smoke that emanates from it. Only the heat release rate and burning rate of the fire and the emission factors for the combustion products need to be specified. The simulation begins several fire diameters downwind of the fire, where the flow field is characterized by relatively small temperature changes, minimal radiation effects, and a velocity field dominated by the prevailing wind. The plume gases ascend to a point in the atmosphere of neutral buoyancy, and then gradually disperse. The trajectory of the plume is governed by the prevailing wind speed, the atmospheric stratification, and the fluctuations of the wind. The wind speed does not vary with height, but the temperature may vary with height according to a prescribed profile.

The uniform wind assumption allows for the transformation of the three dimensional, steady state flow field into a two dimensional, time dependent one. In essence, the fluid motion in the plane normal to the prevailing wind is calculated, starting from the point several fire diameters downwind. The equations and the associated boundary conditions are solved using a relatively simple finite difference technique, and on modern personal computers the calculations typically require a few minutes to complete.

A fixed number of particles are used to represent the combustion products carried aloft and downwind by the plume. The particles are introduced into the flow at the start of the calculation and advected with the induced flow. The initial distribution of particles, like the temperature distribution, is uniform within a circular area in the crosswind plane. The area of the vertical circular area is equal to the fire area and the center is one fire diameter above the surface. All particles move at the same downwind speed equal to that of the wind. The average of the velocity at the start and at the end of a time step is used to update the particle position. Particles that reach the ground level cells are considered to have been deposited on the ground and are not carried forward in the calculation.

Although ALOFT uses large eddy simulation to describe the turbulent motion of the rising smoke plume, it does not directly calculate the turbulence of the atmospheric boundary layer, which may be thought of as variations of the prevailing wind over a time scale of minutes to hours. These deviations are introduced into the model through random perturbations to the

---

[2] NIST is an agency of the U.S. Department of Commerce, and by statute the ALOFT model is not subject to copyright in the United States. However, ALOFT does have trademark protection.

trajectories of the particles. In a smoke plume, the path of particles originating from a given area on the fuel surface changes over time with fluctuations in the wind. Thus, the motion of each particle is governed by the fire-induced velocity field, found by solving the conservation equations of mass, momentum and energy, plus a perturbation velocity field that represents the random temporal and spatial variations of the prevailing wind. The perturbation velocities are determined from the specified standard deviations of the prevailing wind direction in the lateral and vertical directions. This technique of transporting smoke particulate in a fluctuating wind field is similar to techniques used by so-called "puff" models. The only difference is that ALOFT-FT solves the Navier-Stokes equations for the flow field in the transverse plane. The model decouples the random spatial and temporal motion of the advected particles and the governing hydrodynamic equations. The reason for this is that the hydrodynamic equations describe the steady-state plume structure and cannot readily accommodate temporal fluctuations. As a result, use of this model for high wind direction fluctuation values (>25 degrees) is not recommended.

The lateral and vertical fluctuations of the prevailing wind are important input parameters for the ALOFT-FT model. These parameters largely dictate the extent to which the plume expands laterally and vertically once it has risen to its height of neutral buoyancy. The magnitude of the wind fluctuations are correlated with the stability of the atmosphere. Typically, unstable atmospheric conditions (i.e. Pasquill Stability Classes A or B) exhibit greater wind fluctuations than stable conditions (Classes E and F). In the original series of calculations performed with the ALOFT-FT model, Version 1 (McGrattan et al., 1993), it was assumed that the standard deviation of the vertical component of the wind direction, σ_φ, was constant at all elevations. This essentially meant that a parcel of smoke would randomly rise or fall according to the specified vertical wind fluctuation regardless of its height. However, further review of the atmospheric dispersion literature revealed that typically there is less turbulent vertical motion above the mixing height of the atmosphere. For this reason, logic was added to the model that decreased σ_φ above the mixing height by a factor of 2. This change in functionality led to Version 2 of the ALOFT-FT model (McGrattan et al., 1997). The current version of ALOFT-FT is 3.10. The upgrade from version 2 to 3 was due to the packaging of the model as an application for a Windows PC. Versions 1 and 2 did not contain the graphical user interface and plotting program that are included in Version 3. Version 3 does contain the same core solver as version 2.

Validation

The results of three sets of field experiments have been used to assess the accuracy of the ALOFT-FT model. The first, the Newfoundland Offshore Burn Experiment (NOBE), was conducted by Environment Canada in August, 1993 (Walton et al., 1994; Fingas et al., 1995). The second, the Burning of Emulsions Test, was conducted by Alaska Clean Seas (ACS) in September, 1994 (McGrattan et al., 1995). The third set of experiments was a series of diesel fuel burns at the US Coast Guard Fire and Safety Detachment in Mobile, Alabama, conducted in October, 1994 (Walton et al., 1993; McGrattan et al., 1997). For each series of burns, ALOFT-FT was run for the recorded meteorological and burn conditions, and the results were compared with data collected in the field. The model predictions compared favorably with the measured particulate levels, so much so that the safety factor of 2 that was applied to the results of the 1993 modeling study was no longer recommended when the ISB Guidelines were revised in 1997. In all of the field exercises, the ground level PM_{10} concentration, averaged over the duration of the burn, was not observed to exceed 100 $\mu g/m^3$, and the model predictions were typically within the range of variability of the measurements. Because all of the fires lasted on the order of an hour or less, there was considerable spatial and temporal variation in the measured ground level concentrations (see Figure 2).

Model Inputs

The input parameters needed by the ALOFT-FT model and the values used in the current study are listed below. Further discussion of these inputs is included in McGrattan *et al*., 1993.

- Prevailing wind speed (4 m/s, 8 m/s, 12 m/s or 8 knots, 16 knots, 24 knots).
- Standard deviation of the wind direction in the lateral and vertical directions, over both land and water. For the current study, the over land values were used; the same as in the original 1993 study. For Stability Class C and D, the standard deviations of the lateral wind direction are 15° and 10°, respectively. The corresponding values for the vertical direction are 10° and 6°.
- Heat release rate per unit area (see Table 1). For very large hydrocarbon fires, measurements of the thermal radiative output indicate that approximately 90 % of the energy of the fire is entrained into the smoke plume and 10 % is emitted as thermal radiation to the surroundings. Thus, the heat release rate that is input into the model is 90 % of the values given in Table 1.
- Burning rate per unit area (see Table 1).
- Soot yield (see Table 1).
- Temperature profile (see Figure 3).

ALOFT-FT Version History

The ALOFT-FT software has undergone three major revisions. However, the core solver has essentially remained unchanged since version 2. ALOFT-FT version 1 was used to make the original downwind distance predictions for 150 $\mu g/m^3$ (PM_{10}) in 1993 (McGrattan *et al*., 1993). Version 2 was used in the three validation field experiments and is documented in McGrattan *et al*. (1997). There were changes made to the ALOFT-FT algorithm between version 1 and version 2 that would affect the downwind extents of the various particulate levels cited by the different versions of the NAAQS. Version 3 denotes the development of the Windows-based graphical user interface, but it does not include changes to the core solver.

It is important to note that the different versions of ALOFT-FT are not based on changes in the 24-hour NAAQS for particulate. In fact, all versions of ALOFT-FT present hour-averaged particulate concentrations based on the user's choice of PM and concentration level. All versions of ALOFT-FT can draw ground level particulate contours for arbitrary concentrations. The software itself does not specify any particular particulate concentration – users choose the values themselves. The various public release versions (3.x) include routine bug fixes and improvements to the usability of the software. The basic plume algorithm has not changed within the version 3 family, and it remains the same as version 2, the version used in the validation exercises (McGrattan *et al*., 1997).

Version 1: The original program is described in McGrattan *et al*. (1993). At that time, it was referred to as the LES (Large Eddy Simulation) plume model, version 2. LES is the mathematical method used in the model. The use of version 2 was to distinguish the model from a very simple proof of concept code that was cited in an earlier paper. LES plume, version 2, is the model that was referenced in the 1994 ISB Guidelines. This version was simply a Fortran program with no user interface or built-in graphics and was not released to the public. NIST developed other fire models based on LES and to distinguish the models from one another, the outdoor smoke plume trajectory model was named ALOFT (A Large Outdoor Fire plume Trajectory model). The original ALOFT was for flat terrain and was later referred to as ALOFT-FT, to distinguish it from a complex terrain version, ALOFT-CT[3].

Version 2: The basic algorithm was modified, primarily to change the assumption of equal vertical mixing at all elevations (see discussion of the Mathematical Model above). This version was used in the three validation exercises described above. In general, ALOFT-FT version 2 predicts lower ground level particulate concentrations than version 1, primarily because it assumes less vertical mixing of the plume above the mixing height of the atmosphere. Version 2 was also a Fortran-only program and was not released to the public. Neither version 1 nor version 2 was referred to as ALOFT-FT when originally published.

Version 3.0: The Fortran source code was packaged along with a graphical user interface and post-processing package to become officially known as ALOFT-FT, version 3.0. It uses the same basic algorithm as version 2. This is the first version of ALOFT-FT released to the public in 1997.

Version 3.1: A new version of the Fortran complier with improved precision was used which resulted in slight differences in the output. Minor programming bugs were fixed, but these did not change the results. The ALOFT-FT trademark was registered.

Version 3.2: Minor bug fixes were made that did not change the results. This is the version used for the present report.

ALOFT-CT (Complex Terrain)

The original ALOFT model was developed under the assumption that the smoke plume was primarily lofted over water or a relatively flat shoreline. The meaning of "flat" is that the change in terrain height is not more than 10 % of the lofting height of the plume. This assumption is based primarily on the fact that the ALOFT-FT (Flat Terrain) model solves a simplified set of the governing fluid flow equations in which the prevailing wind speed is constant and that there are no terrain-induced air currents to increase or decrease the height of the plume once aloft. The ALOFT-CT (Complex Terrain) model added a three-dimensional wind field and allowed the smoke to follow the contours of complex terrain.

The ALOFT-FT model, version 1, was used to develop the original ISB Guidelines in 1994. These Guidelines were based on simulations of hypothetical fires of various sizes under a variety

---

[3] A version of ALOFT for mountainous or complex terrain, ALOFT-CT, was developed in 1995. This model was for research purposes only and was not released to the public. Results of its calculations are reported in McGrattan *et al*. (1997).

of weather conditions typical of both the North Slope and Cook Inlet. Following this study, the ALOFT-CT model was developed with support from ADEC and the U.S. Minerals Management Service (McGrattan *et al.*, 1997). The ALOFT-CT model was used to predict the downwind concentration of particulate from hypothetical fires in different regions of Alaska, both on the coast and in the interior (along the Alyeska Pipeline).

The ALOFT-CT model was never experimentally validated. The three series of experiments cited above were all conducted over relatively flat terrain, and the measurements were compared against predictions made by ALOFT-FT. For this reason, ALOFT-CT was not used to develop ISB Guidelines. It was, and remains, what is considered a "research code." It was decided by the sponsoring organizations not to develop an easy to use Windows-based graphical user interface for the CT version like the one developed for the FT version. The CT version requires far more computational resources than the FT version. It would be difficult, if not impossible, to forecast smoke plume trajectories for every conceivable spill location in and around Alaska because of the variety of terrain and weather conditions. It would be equally difficult to run the ALOFT-CT model during an actual spill event because of the time and computer resources needed.

ALOFT-FT Modeling Results

This section summarizes the results of updated calculations that predict the downwind ground level concentrations of smoke particulate from a variety of hypothetical in situ burns of oil. These modeling results were first presented in McGrattan, *et al.* (1993), but they have been updated here to reflect the latest NAAQS particulate standard. In 1993, the 24-hour average standard for PM_{10} was 150 $\mu g/m^3$, whereas the current standard also includes an additional 24-hour average for $PM_{2.5}$ of 35 $\mu g/m^3$.

Following are some important points to consider:

- The original 1993 ALOFT-FT calculations, which subsequently were used in the 1994 ISB Guidelines, were performed with ALOFT-FT version 1. It was not labeled as such at the time because the acronym ALOFT-FT was not adopted until 1997 when the model was packaged for public release.

- The results of the calculations, presented in Table 3, were performed with ALOFT-FT version 3.20. This version has the same basic plume trajectory algorithm as ALOFT-FT version 2, the version that was validated against the three sets of field experiments.

- It is assumed in the model that the specified wind fluctuations are appropriate for a 1-hour average prediction of ground level particulate because it is assumed that *in situ* burns of spilled oil typically last for a few hours[4]. The model assumes that the hourly-averaged wind fluctuations are correlated with various atmospheric "stability classes."

- The downwind extent of the various $PM_{2.5}$ concentration contours (see Appendix) are determined by visual inspection. Typically, the extent of a given concentration is where there is no longer a continuous pattern.

---

[4] During the *Deepwater Horizon* burns of 2010, the burn time ranged from approximately 30 min to 12 h

Table 2. Summary of fire scenarios.

Fire Area m^2 (ft^2)	232 (2500)		465 (5000)		930 (10000)	
Region	Cook Inlet	North Slope	Cook Inlet	North Slope	Cook Inlet	North Slope
Total Heat Release Rate (MW)	505	455	1010	910	2030	1820
Convective Heat Release Rate (MW)	455	410	900	820	1825	1640
Burning Rate (kg/s)	13.0	11.8	26.0	23.7	52.1	47.4
Smoke (PM$_{2.5}$) Release Rate (kg/s)	0.832	1.01	1.66	2.04	3.33	4.08

Table 3. Predicted extents of three different ground level hour-average concentrations of smoke particulate from a variety of fires under different weather conditions.

Location	Season	Stability Class	Wind Speed (m/s)	Extent of 25 µg/m³ (km)	Extent of 35 µg/m³ (km)	Extent of 65 µg/m³ (km)
Burning Area of 232 m² (2500 ft²)						
Cook Inlet	Summer	C	4	7	5	3
Cook Inlet	Summer	D	8	6	3	2
Cook Inlet	Summer	D	12	5	4	3
Cook Inlet	Winter	C	4	<1	<1	<1
Cook Inlet	Winter	D	8	<1	<1	<1
Cook Inlet	Winter	D	12	3	2	1
North Slope	Summer	C	4	<1	<1	<1
North Slope	Summer	D	8	<1	<1	<1
North Slope	Summer	D	12	1	1	1
North Slope	Winter	C	4	<1	<1	<1
North Slope	Winter	D	8	<1	<1	<1
North Slope	Winter	D	12	2	2	1
Burning Area of 465 m² (5000 ft²)						
Cook Inlet	Summer	C	4	10	8	<1
Cook Inlet	Summer	D	8	10	8	<1
Cook Inlet	Summer	D	12	8	5	4
Cook Inlet	Summer	C	4	<1	<1	<1
Cook Inlet	Summer	D	8	<1	<1	<1
Cook Inlet	Summer	D	12	<1	<1	<1
North Slope	Summer	C	4	<1	<1	<1
North Slope	Summer	D	8	<1	<1	<1
North Slope	Summer	D	12	<1	<1	<1
North Slope	Summer	C	4	<1	<1	<1
North Slope	Summer	D	8	<1	<1	<1
North Slope	Summer	D	12	<1	<1	<1
Burning Area of 930 m² (10000 ft²)						
Cook Inlet	Summer	D	8	10	10	<1
Cook Inlet	Winter	D	8	<1	<1	<1
North Slope	Summer	D	8	<1	<1	<1
North Slope	Winter	D	8	<1	<1	<1

Discussion of ALOFT-FT Results

The model results shown in Table 3 are predictions of the downwind extents of 1-hour average ground level $PM_{2.5}$ concentrations of 65 $\mu g/m^3$, 35 $\mu g/m^3$, and 25 $\mu g/m^3$. Table 3 is analogous to Table 9 of McGrattan *et al.*, 1993, which presented predictions of the extent of ground level PM_{10} concentrations of 150 $\mu g/m^3$. The 1993 report concluded:

> For the conditions considered in the report and summarized in Table 9, we found that the LES plume trajectory model [ALOFT-FT version 1] predicts that hour-averaged ground level particulate concentrations of 150 $\mu g/m^3$ or higher do not extend beyond the first 5 kilometers downwind of the burn site, nor do these levels extend outside a path of about a kilometer in width.

A similar examination of the predicted downwind concentrations for the same 28 scenarios considered in the 1993 study indicates that hour-averaged ground level particulate concentrations of 65 $\mu g/m^3$ ($PM_{2.5}$) do not extend beyond the 4 km (2.5 mi). Concentrations of hour-averaged 25 $\mu g/m^3$ and 35 $\mu g/m^3$ ($PM_{2.5}$) do not extend beyond 10 km (6 mi).

The 1993 calculations were different than the current set in two respects. First, the production rate of $PM_{2.5}$ is approximately 70 % that of PM_{10} (McGrattan *et al.*, 1993). Thus, the 1993 predictions were made with a larger effective production rate of smoke particulate, a parameter that would tend to increase the extent of the 150 $\mu g/m^3$ contour further than that of $PM_{2.5}$ results. Second, the 1993 predictions were made with ALOFT-FT version 1. This version of the model assumed uniform vertical mixing at all heights of the atmosphere, an assumption that was not supported by the validation experiments and an assumption that tended to predict higher ground level particulate concentrations than predictions of ALOFT-FT versions 2 and 3. Thus, the 1993 results tend to show higher ground level concentrations of particulate, regardless of its size, than the current results.

The ground level concentrations included in Table 3 (65 $\mu g/m^3$, 35 $\mu g/m^3$, and 25 $\mu g/m^3$) represent different values of the national ambient air particulate standards, except that the short term NAAQS for particulate is a 24 hour, not 1 hour, average. It should be understood that these concentrations of smoke[5] are difficult to measure or predict on a short term basis. Typically, air quality models and measurement techniques apply time-averaging or air sampling over much longer periods of time than an hour. Because of this, the uncertainty of the predictions should be noted. There are two forms of uncertainty in any model prediction – uncertainty in the input parameters and uncertainty in the mathematical model itself. These are usually referred to as *parameter uncertainty* and *model uncertainty*.

To address the parameter uncertainty of the model, the ARRT decided in 1994 that it would adopt one set of "safe distances" for ISB rather than attempt to predict (or *forecast*) potential ground level particulate concentrations on the day of an actual burn. Although it is never possible to formulate a "worst case" set of conditions that would give rise to the highest possible

---

[5] During the Alaska Clean Seas Burning of Emulsions Experiment in 1994, the people measuring the ground level smoke with real time aerosol monitors (RAMs) noted that it was difficult to either see or smell the smoke at concentrations less than approximately 100 $\mu g/m^3$.

ground level concentrations, the results of the ALOFT-FT model runs included in the Appendix provide a reasonable sampling of possible outcomes for a wide variety of conditions. This strategy is often used in fire protection engineering. Fire safety requirements are often based on the most severe outcome of a large sample of potential fire scenarios because it would be impractical to design to, or apply any kind of model to every possible scenario. In taking this approach, the regulatory authority acknowledges that there is uncertainty associated with model inputs and bases its decision making on a relatively severe rather than typical fire scenario.

It is even more difficult to quantify the model uncertainty. Typically, the uncertainty of a fire model is quantified by comparing its predictions to full-scale experiments. The best set of full-scale experiments with which to compare ALOFT-FT predictions is the ACS Burning of Emulsions Experiment, September, 1994 (McGrattan *et al.*, 1995). The result of the comparison is shown in Figure 2. The ALOFT-FT model captures the general trend and magnitude of the measurements, but it is clear that it cannot predict precisely the smoke concentration at every measurement location. A good analogy to this is weather prediction. A current generation weather forecast is reasonably accurate in predicting the likelihood of thunderstorms for a particular region, but it cannot predict whether or not those storms are likely to impact a given neighborhood. Over the course of hours, meteorological conditions vary in both space and time, and the ALOFT-FT model predictions reflect this because they employ randomness in the transport of the particulate matter through a predetermined wind field. For example, the result of ACS Burn 1 in Figure 2 shows a measured PM_{10} concentration of 85 $\mu g/m^3$ at a distance of 4 km from the fire, whereas neither the model nor the surrounding measurements show values nearly that high at this distance downwind. Examination of the data (McGrattan *et al.*, 1995) reveals that a relatively high concentration of smoke was observed towards the end of the experiment when the fire was nearly extinguished. During these transient stages of the fire, it is not unusual to see smoke closer to the ground.

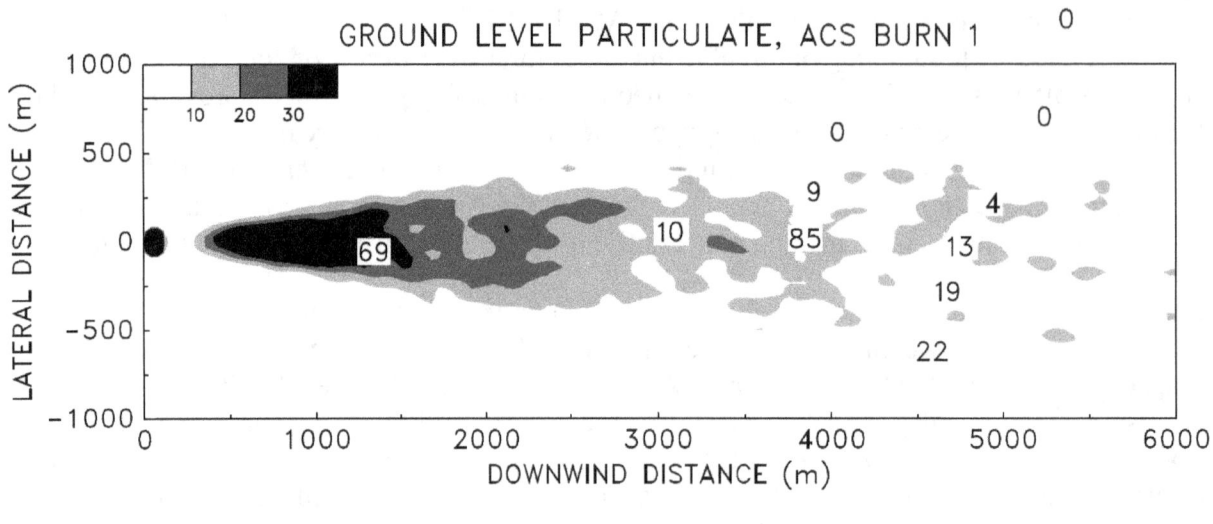

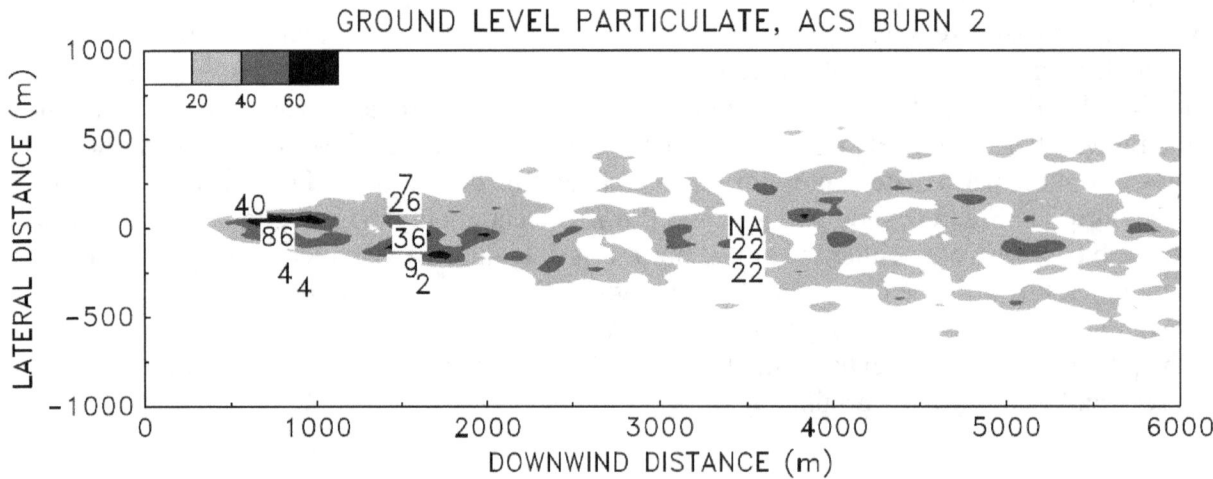

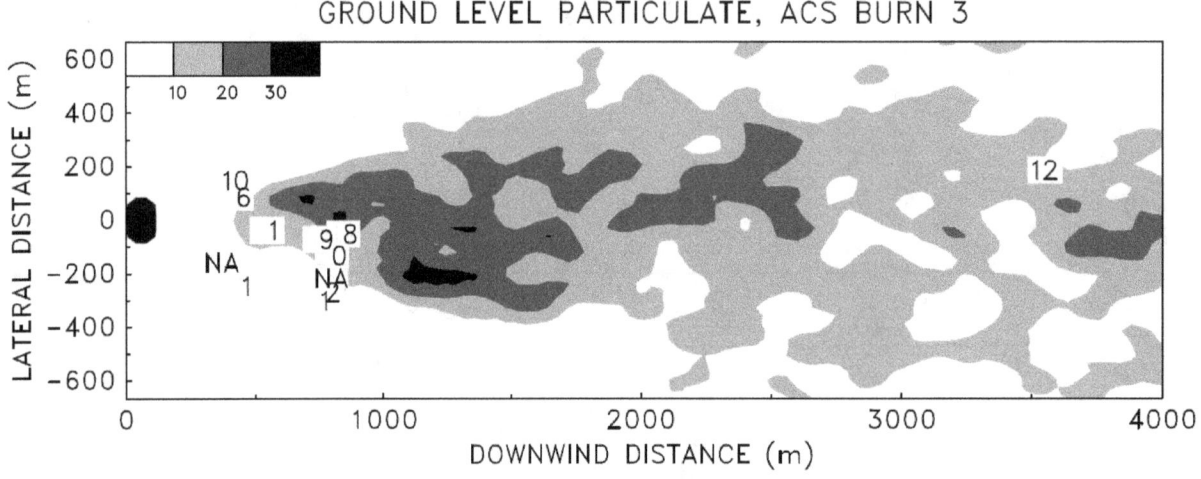

Figure 2. Comparison of ALOFT-FT predictions (shaded contours) with field measurements (numbers) for the ACS Burning of Emulsions Experiment. Units are µg/m³.

18

Conclusion

For nearly 20 years, the ALOFT (A Large Outdoor Fire plume Trajectory) model has been used to predict the concentration of smoke particulate from *in situ* burns of oil. The calculations have been used for the development of *in situ* burning guidelines in Alaska and elsewhere. In particular, the ALOFT-FT predictions have been used to determine "safe distances" for populated areas that may be downwind of a potential burn site. Model predictions were first made in 1993 and contributed to the first ISB Guidelines in Alaska in 1994. In the years since, three sets of validation experiments were conducted in Newfoundland, Canada, Prudhoe Bay, Alaska, and Mobile, Alabama, that further added validity to the model predictions. The model was released for public use and it was subsequently used by other Regional Response Teams (RRTs) in the development of ISB Guidelines similar to those developed in Alaska. Several *in situ* burns have been conducted to remediate actual oil spills, most notably following the loss of the *Deepwater Horizon* platform in 2010. An examination of the burn sizes and duration during this incident revealed that the size estimates in the original 1993 modeling study are typical of actual *in situ* burning operations.

The purpose of the current report is to update the original 1993 ALOFT-FT predictions in light of changes to the National Ambient Air Quality Standards. The 1993 24-hour PM_{10} standard for was 150 $\mu g/m^3$, followed in 1997 by an additional 24-hour $PM_{2.5}$ standard of 65 $\mu g/m^3$, which in 2006 was reduced to 35 $\mu g/m^3$. Using the most recent version of the ALOFT-FT model, the extent of hour-averaged $PM_{2.5}$ concentrations of 25 $\mu g/m^3$ and 35 $\mu g/m^3$ is predicted to be within 10 km of the fire. According to the original version of the model in 1993, the predicted extent of the hour-averaged PM_{10} concentration of 150 $\mu g/m^3$ was 5 km.

References

Allen, A. A., "New Tools and Techniques for Controlled In-Situ Burning," *Proceedings of the 22nd Arctic and Marine Oilspill Program (AMOP) Technical Seminar*, Volume 2, June 2-4, 1999, Calgary, Alberta, Canada, Environment Canada, Ottawa, Ontario, pp. 613-628, 1999.

Allen, A. A., "Controlled Offshore Burning of Spills," Clean Gulf Pre-Conference Workshop, October 18, 2010, Tampa, Florida.

Evans, D.D., G.W. Mulholland, H.R. Baum, W.D. Walton and K.B. McGrattan, "In Situ Burning of Oil Spills," *Journal of Research of the National Institute of Standards and Technology*, Vol. 106, No. 1, pp. 231-278, 2001.

Fingas, M.F. *et al.*, The Newfoundland Offshore Burn Experiment – NOBE," Proceedings of the 1995 International Oil Spill Conference, Long Beach, California, Feb. 27 – Mar. 2, 1995, American Petroleum Institute Pub. No. 4620, pp 123-132.

Lehr, W., S. Bristol, and A. Possolo, *Oil Budget Calculator, Deepwater Horizon, A Report by the Federal Interagency Solutions Group to the National Incident Command*, National Oceanic and Atmospheric Administration, November, 2010.

McGrattan, K.B., A.D. Putorti, W.H. Twilley and D.D. Evans, "Smoke Plume Trajectory of *In Situ* Burning of Crude Oil in Alaska", NISTIR 5273, National Institute of Standards and Technology, Gaithersburg, Maryland, 20899, October, 1993.

McGrattan, K.B., W.D. Walton, A.D. Putorti, W.H. Twilley, J. McElroy and D.D. Evans, "Smoke Plume Trajectory of *In Situ* Burning of Crude Oil in Alaska: Field Experiments", NISTIR 5764, National Institute of Standards and Technology, Gaithersburg, Maryland, 20899, November, 1995.

McGrattan, K.B., H.R. Baum, W.D. Walton, and J. Trelles, "Smoke Plume Trajectory of *In Situ* Burning of Crude Oil in Alaska – Field Experiments and Modeling of Complex Terrain", NISTIR 5958, National Institute of Standards and Technology, Gaithersburg, Maryland, 20899, January, 1997.

Pope, C.A., R.T. Burnett, M.J. Thun, E.E. Calle, D. Krewski, K. Ito and G.D. Thurston, "Lung Cancer, Cardiopulmonary Mortality, and Long-term Exposure to Fine Particulate Air Pollution," *J. American Medical Association*, Vol. 287, No. 9, pp. 1132-1141, March, 2002.

Pope, C.A., J.B. Muhlestein, H.T. May, D.G. Renlund, J.L. Anderson, B.D. Horne, "Ischemic Heart Disease Events Triggered by Short-Term Exposure to Fine Particulate Air Pollution," *Circulation*, Vol. 114, pp. 2443-2448, 2006.

Walton, W.D. *et al.*, "In Situ Burning of Oil Spills: Meso-scale Experiments and Analysis," Proceedings of the Sixteenth Arctic and Marine Oil Spill Program Technical Seminar, June 7-9, 1993, Calgary, Alberta, pp. 679-734.

Walton, W.D. *et al.*, "Smoke Measurements Using a Tethered Miniblimp at the Newfoundland Offshore Oil Burn Experiment," Proceedings of the Seventeenth Arctic and Marine Oil Spill Program Technical Seminar, June 8-10, 1994, Vancouver, British Columbia, pp. 1083-1098.

Appendix

This appendix contains the results of the 28 ALOFT-FT simulations listed in Table 3. There are 28 cases in all, including 3 different fire sizes, 3 different wind speeds, and 4 different temperature profiles. The 4 temperature profiles, shown in Figure 3, are labeled with a region in Alaska and a season for which this type of profile is typical.

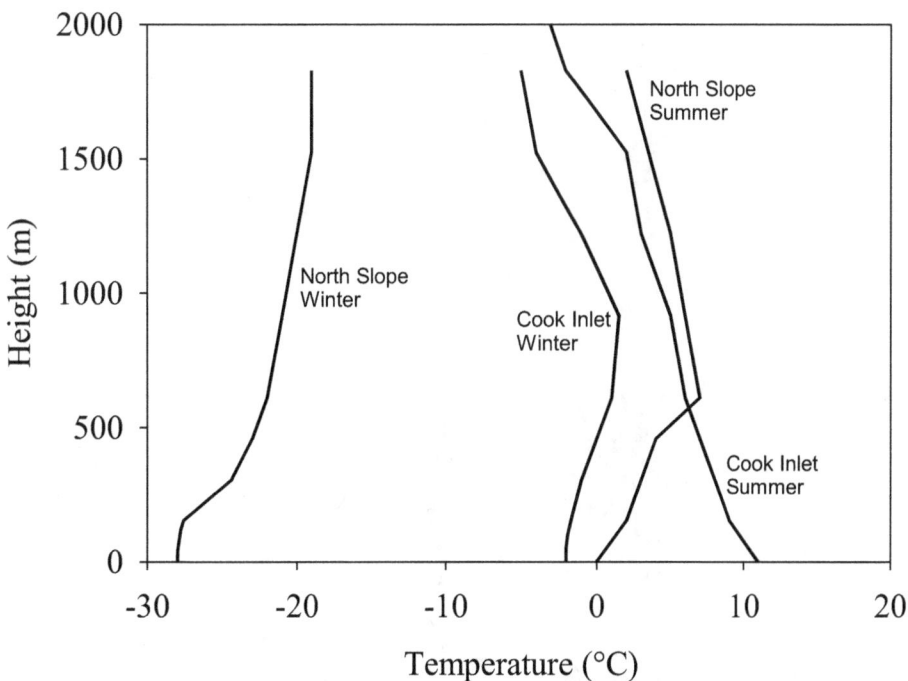

Figure 3. Temperature profiles for the four meteorological conditions (McGrattan *et al.*, 1993).

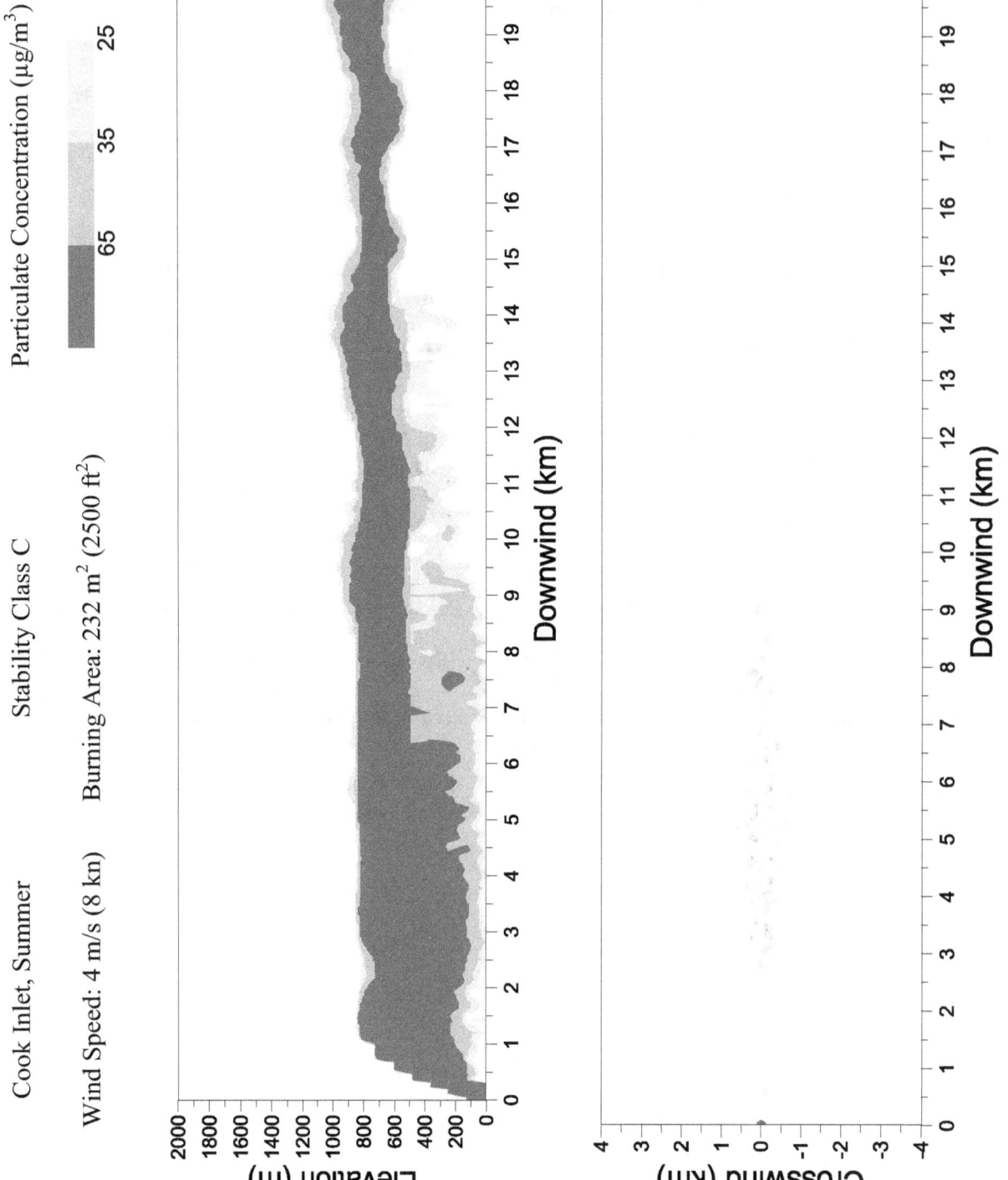

Cook Inlet, Summer   Stability Class C   Particulate Concentration (μg/m³)

Wind Speed: 4 m/s (8 kn)   Burning Area: 232 m² (2500 ft²)

65   35   25

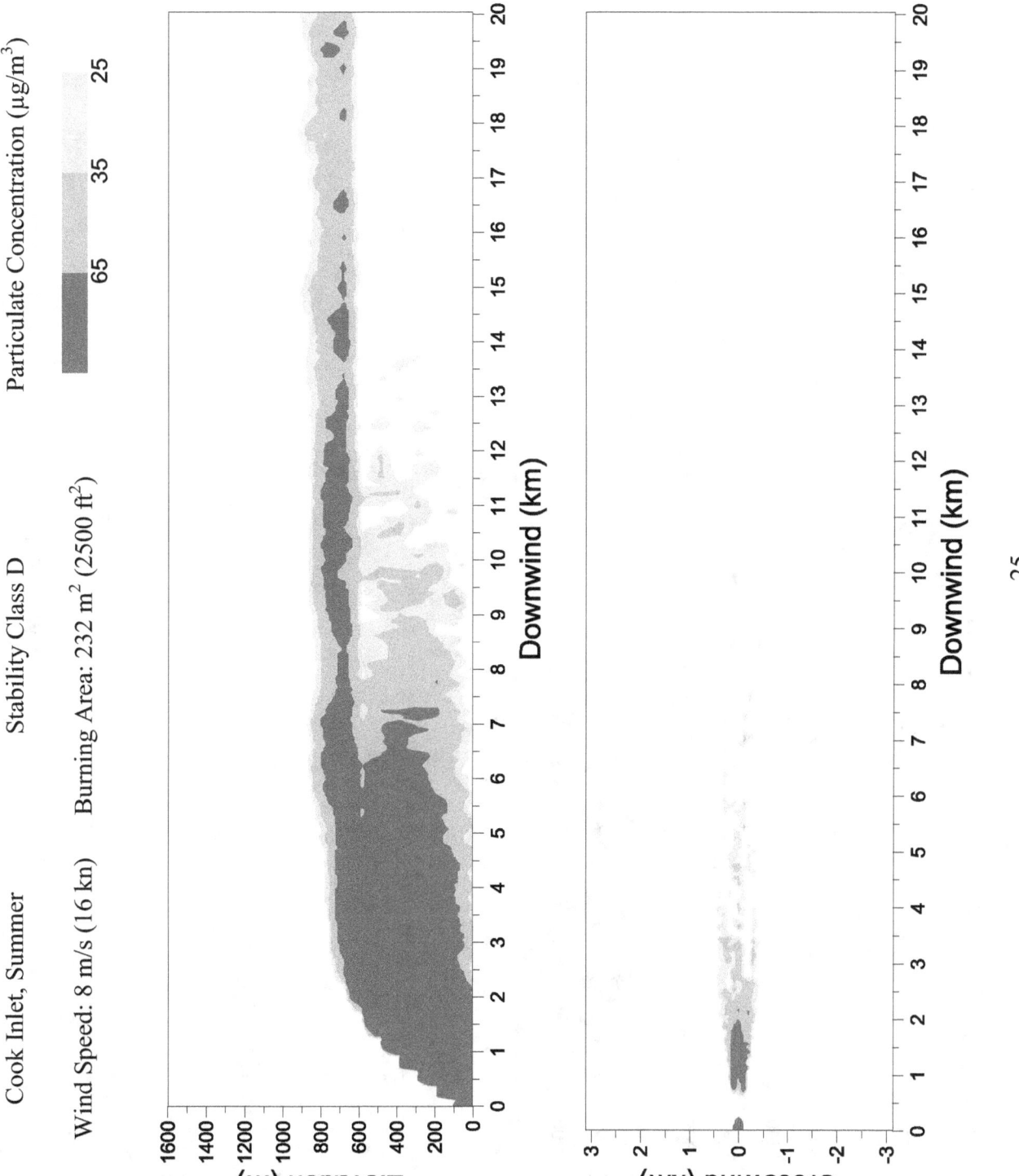

Cook Inlet, Summer   Stability Class D   Particulate Concentration (μg/m³)

Wind Speed: 8 m/s (16 kn)   Burning Area: 232 m² (2500 ft²)

65   35   25

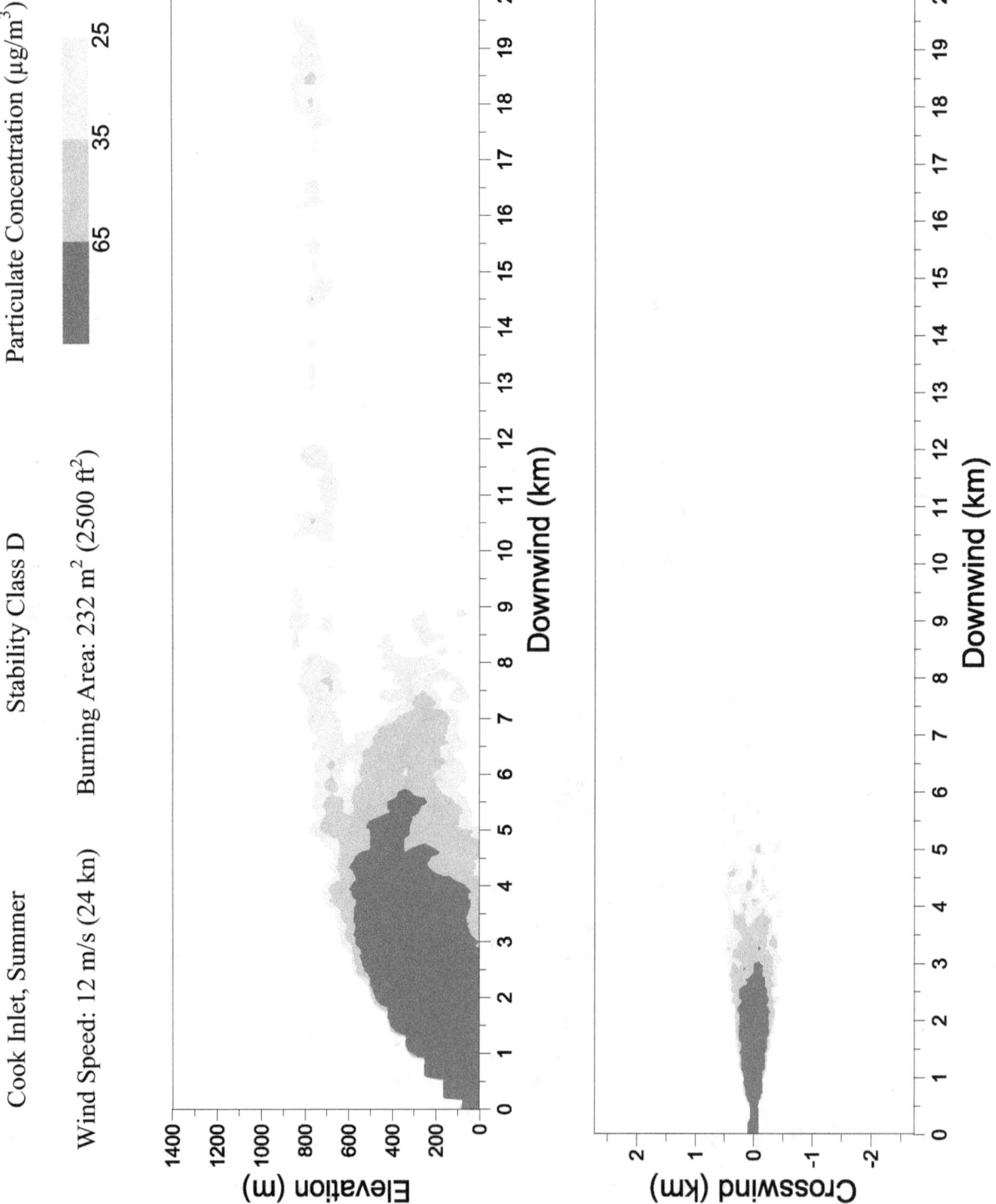

Cook Inlet, Summer   Stability Class D   Particulate Concentration (µg/m³)

Wind Speed: 12 m/s (24 kn)   Burning Area: 232 m² (2500 ft²)

25   35   65

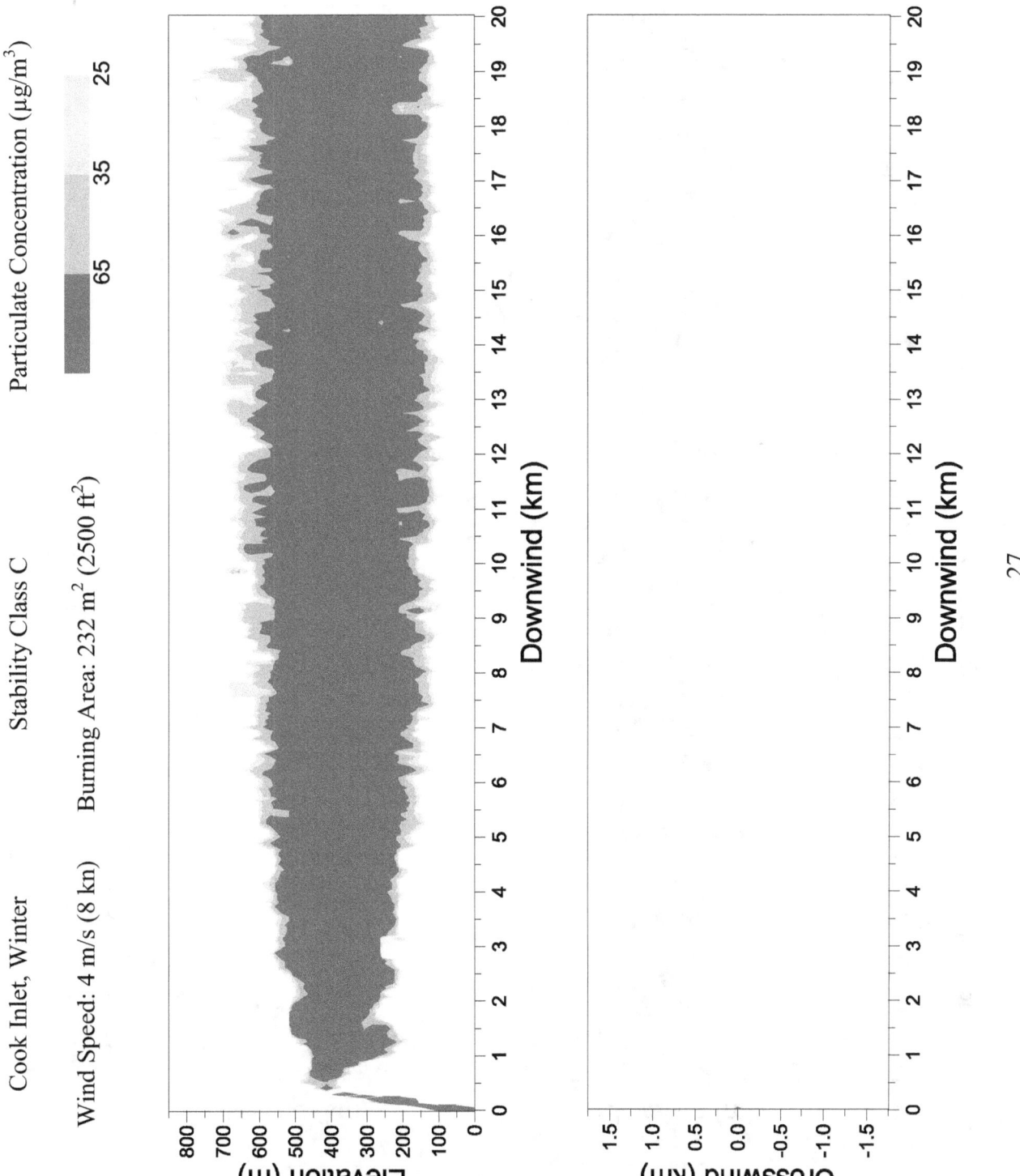

Cook Inlet, Winter   Stability Class C   Particulate Concentration ($\mu g/m^3$)

Wind Speed: 4 m/s (8 kn)   Burning Area: 232 m^2 (2500 ft^2)

27

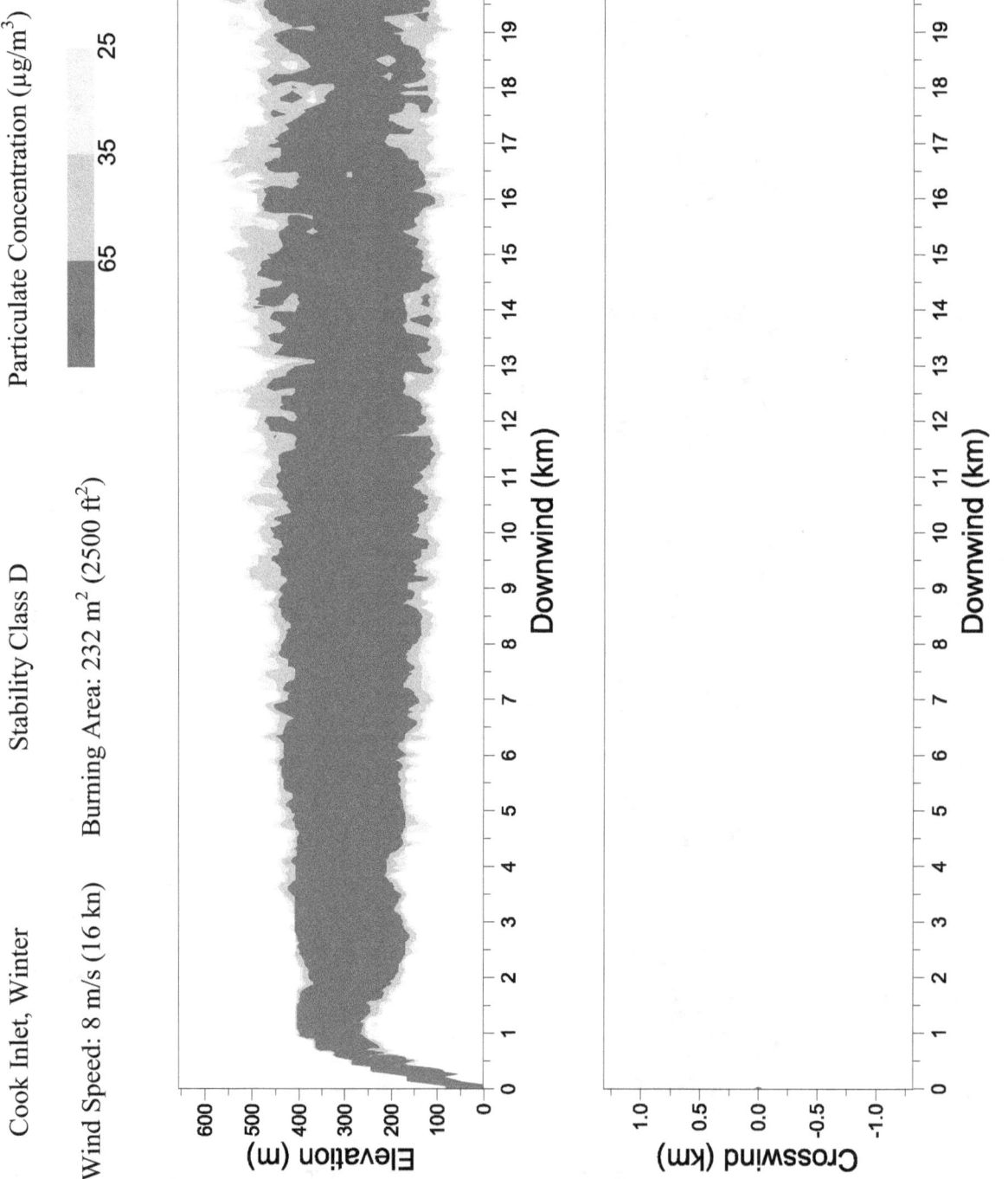

Cook Inlet, Winter

Stability Class D

Particulate Concentration ($\mu g/m^3$)

Wind Speed: 8 m/s (16 kn)   Burning Area: 232 m^2 (2500 ft^2)

25   35   65

Cook Inlet, Winter   Stability Class D   Particulate Concentration (µg/m^3)

Wind Speed: 12 m/s (24 kn)   Burning Area: 232 m^2 (2500 ft^2)

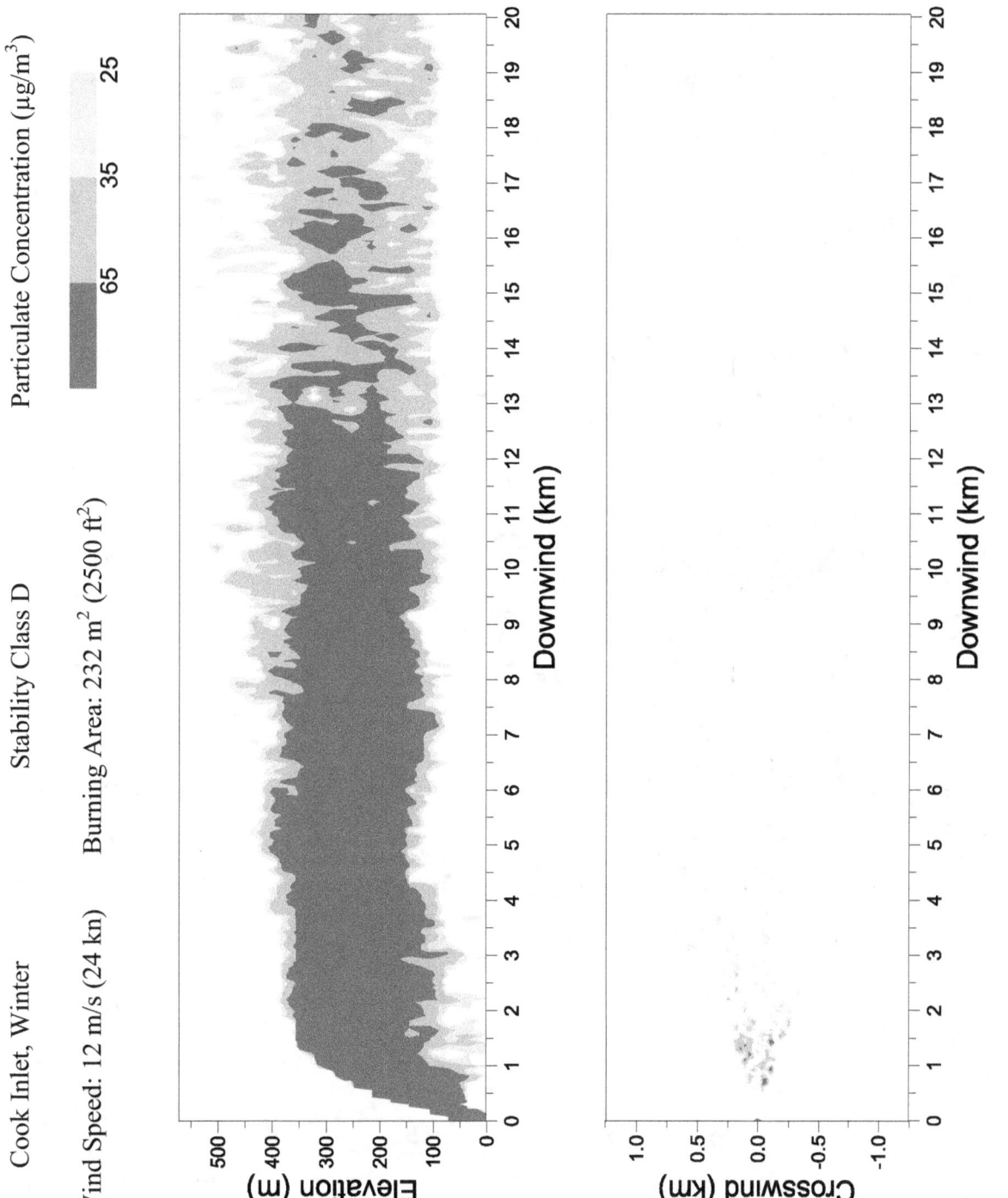

29

North Slope, Summer   Stability Class C

Wind Speed: 4 m/s (8 kn)   Burning Area: 232 m² (2500 ft²)

Particulate Concentration (µg/m³)

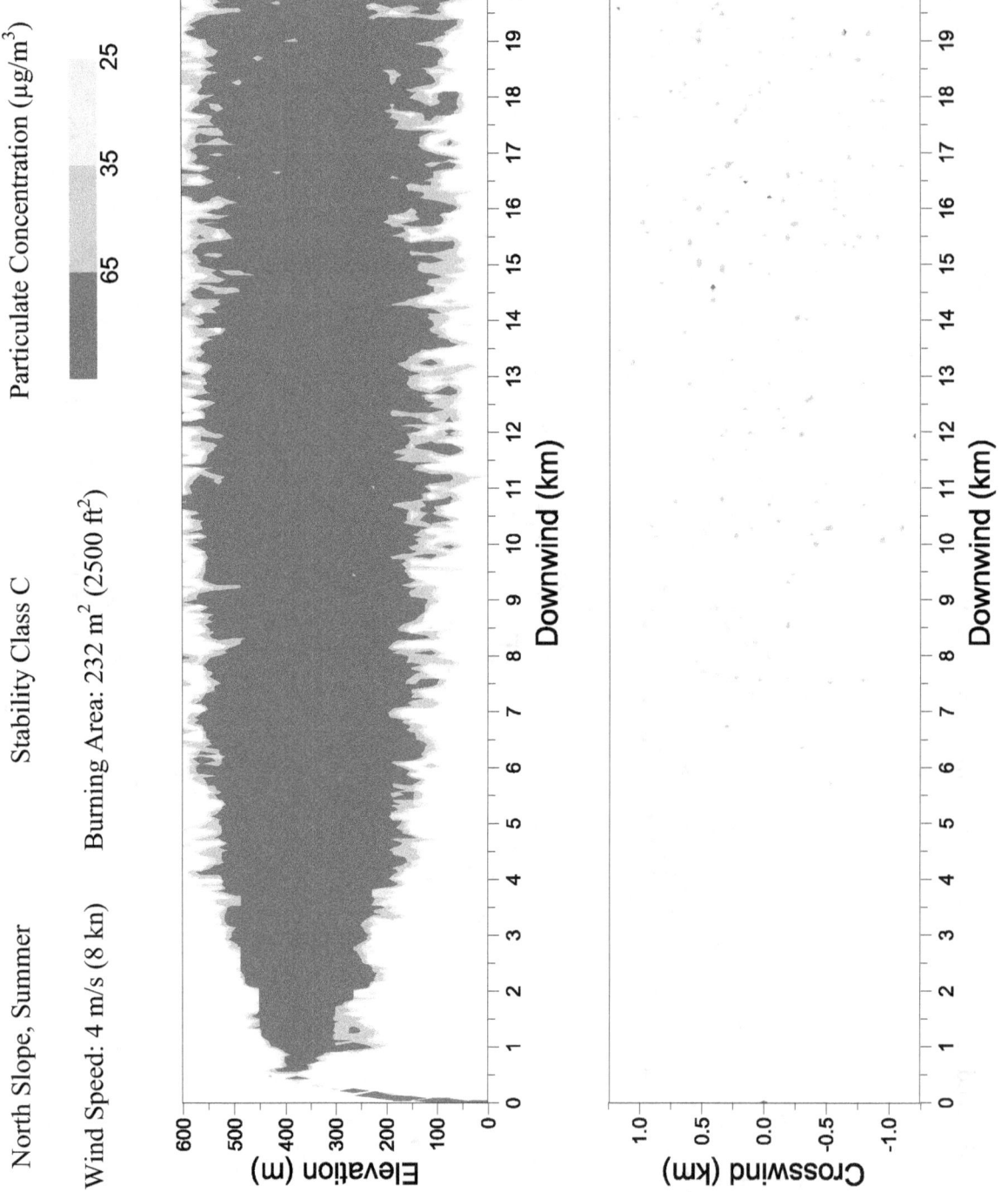

North Slope, Summer   Stability Class D   Particulate Concentration (µg/m³)

Wind Speed: 8 m/s (16 kn)   Burning Area: 232 m² (2500 ft²)

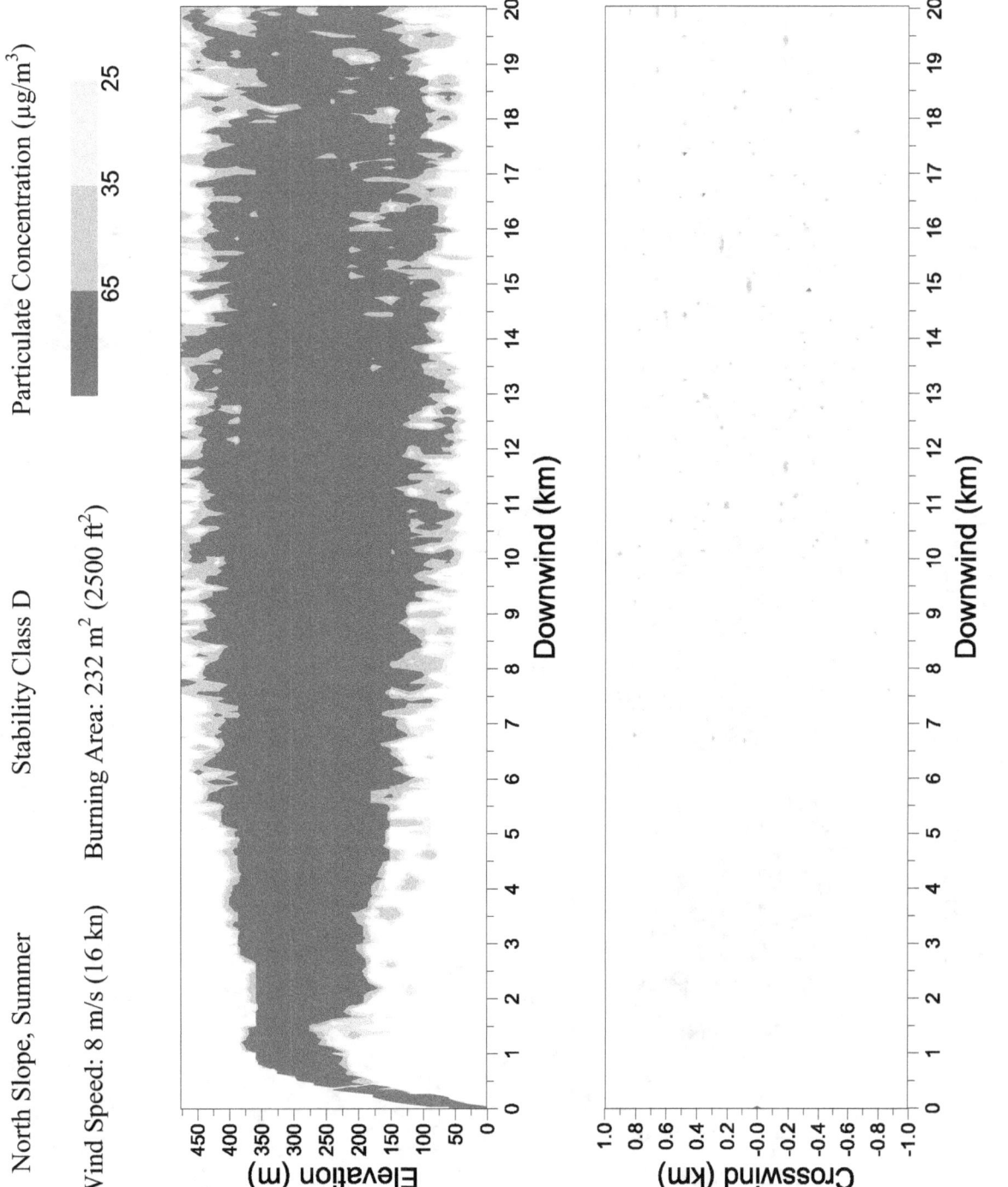

North Slope, Summer   Stability Class D   Particulate Concentration ($\mu g/m^3$)

Wind Speed: 12 m/s (24 kn)   Burning Area: 232 m^2 (2500 ft^2)

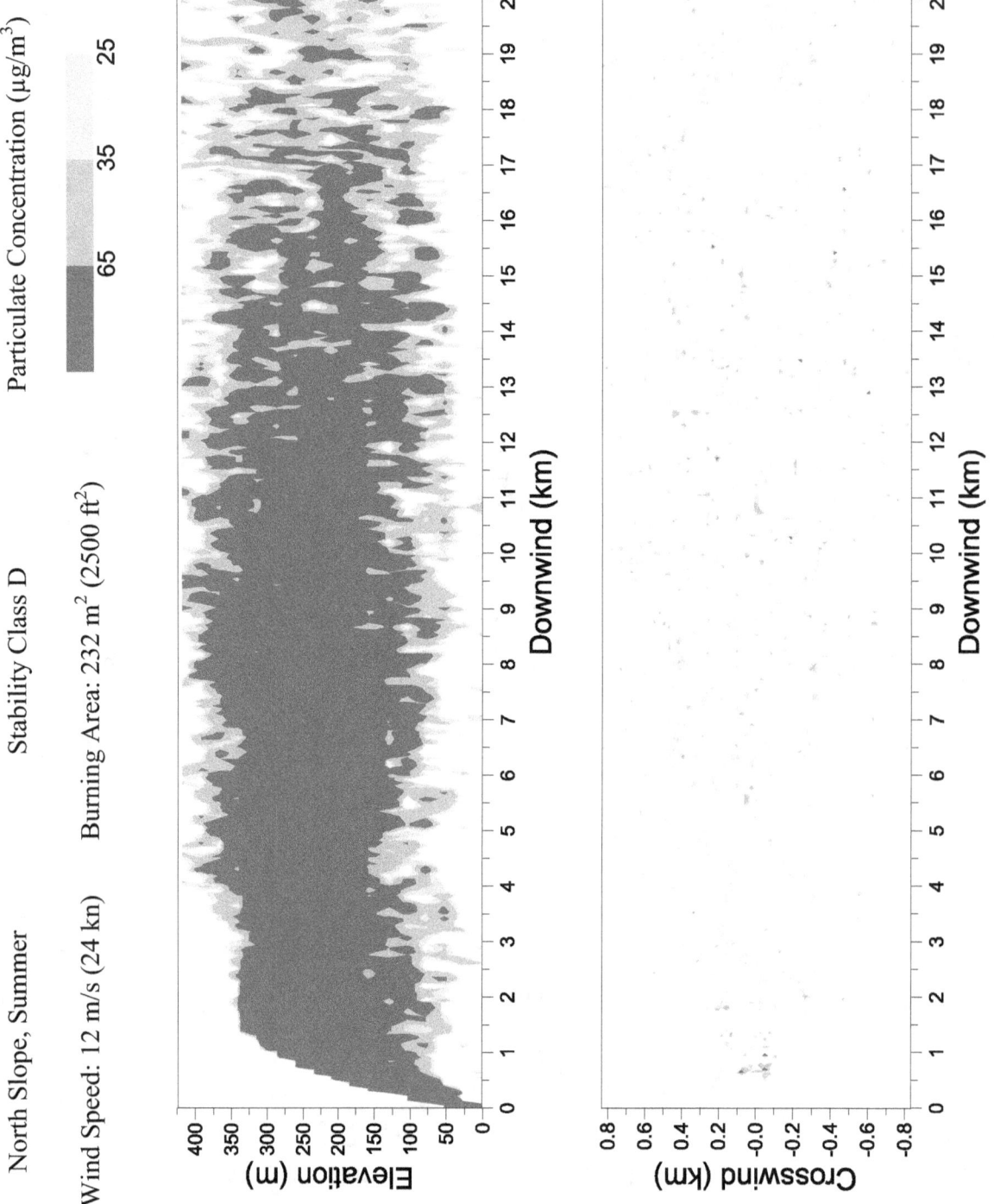

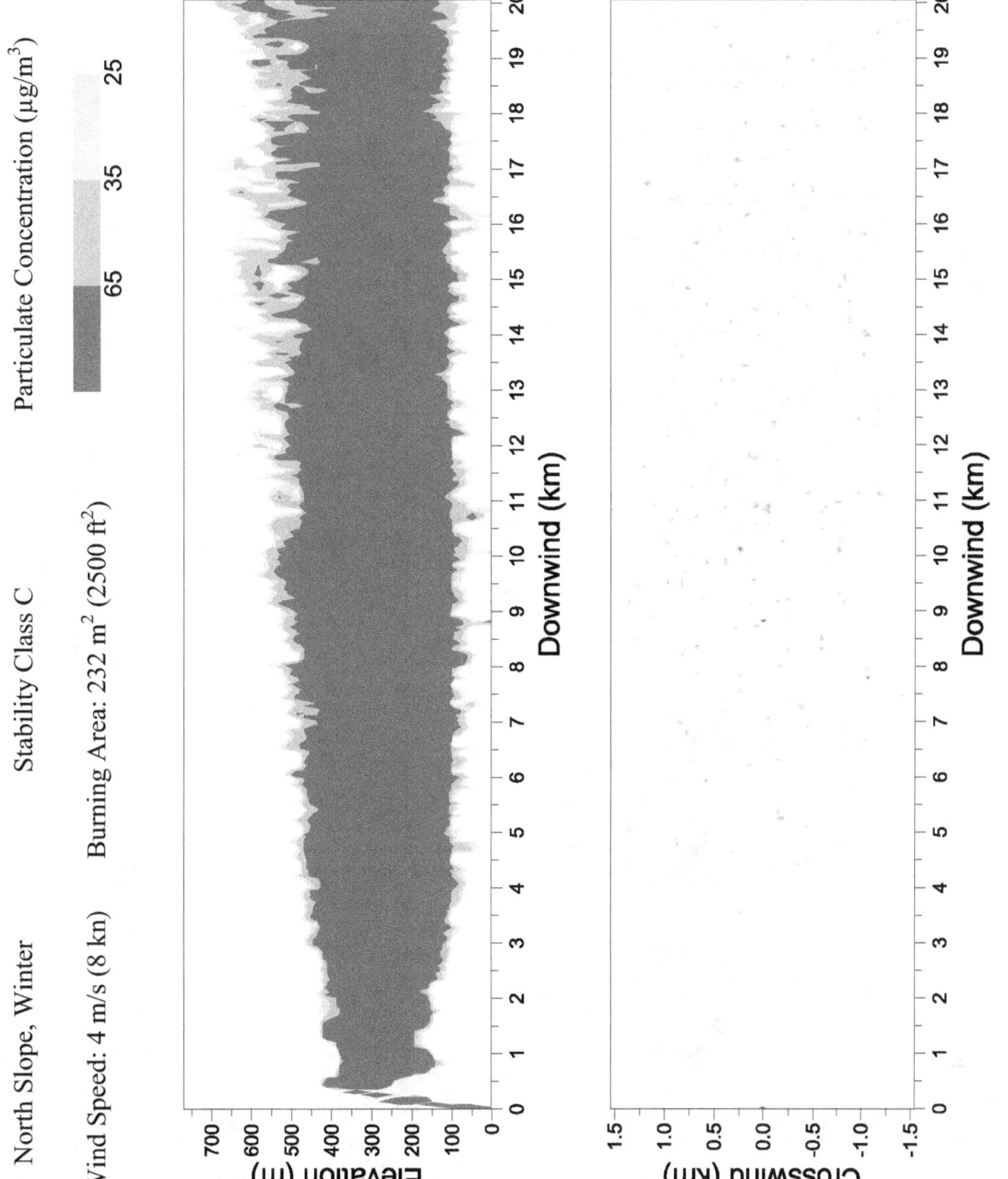

North Slope, Winter   Stability Class C   Particulate Concentration (µg/m³)

Wind Speed: 4 m/s (8 kn)   Burning Area: 232 m² (2500 ft²)

25   35   65

33

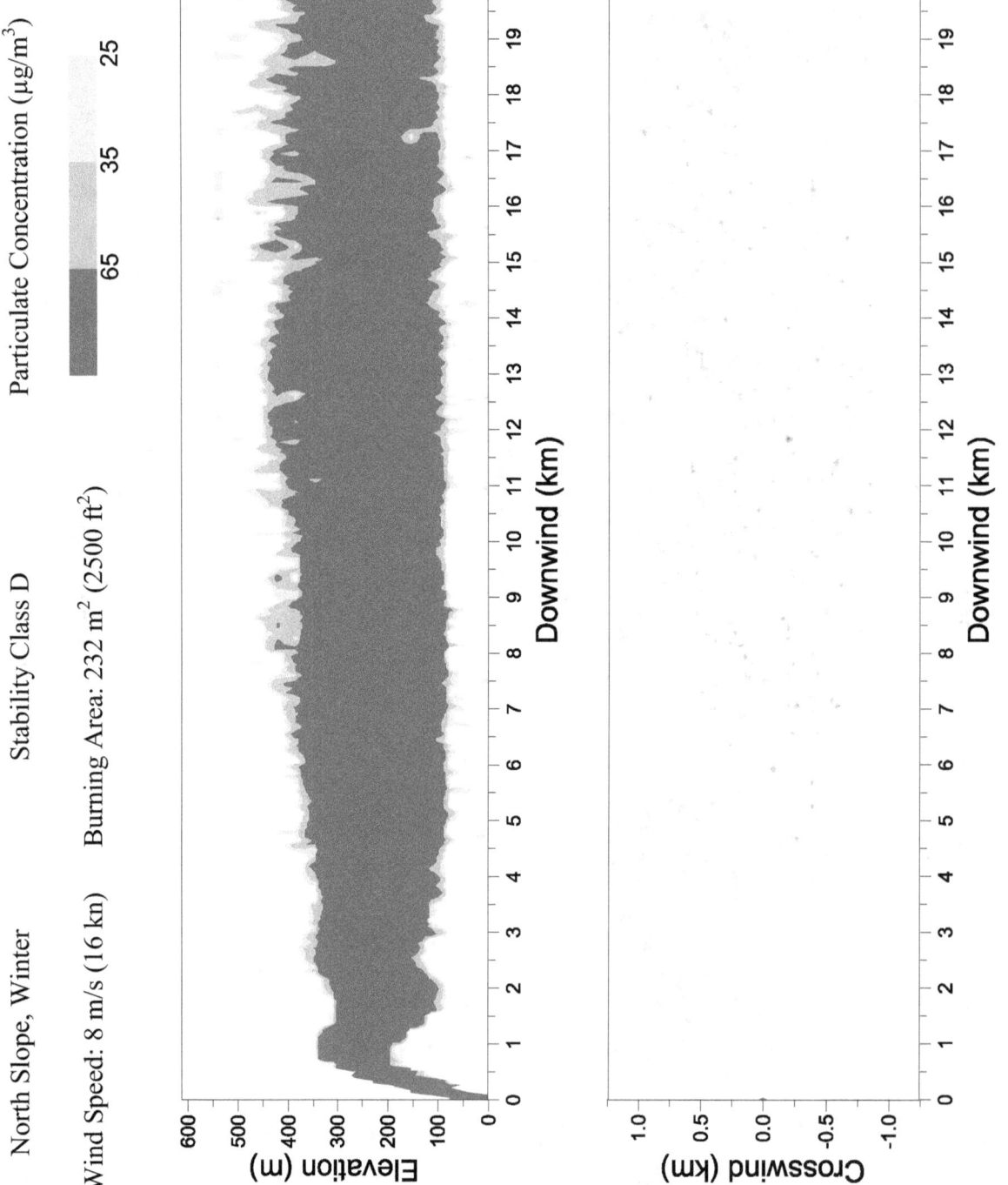

North Slope, Winter   Stability Class D   Particulate Concentration (μg/m³)

Wind Speed: 8 m/s (16 kn)   Burning Area: 232 m² (2500 ft²)

65   35   25

34

North Slope, Winter

Stability Class D

Particulate Concentration (µg/m³)

Wind Speed: 12 m/s (24 kn)

Burning Area: 232 m² (2500 ft²)

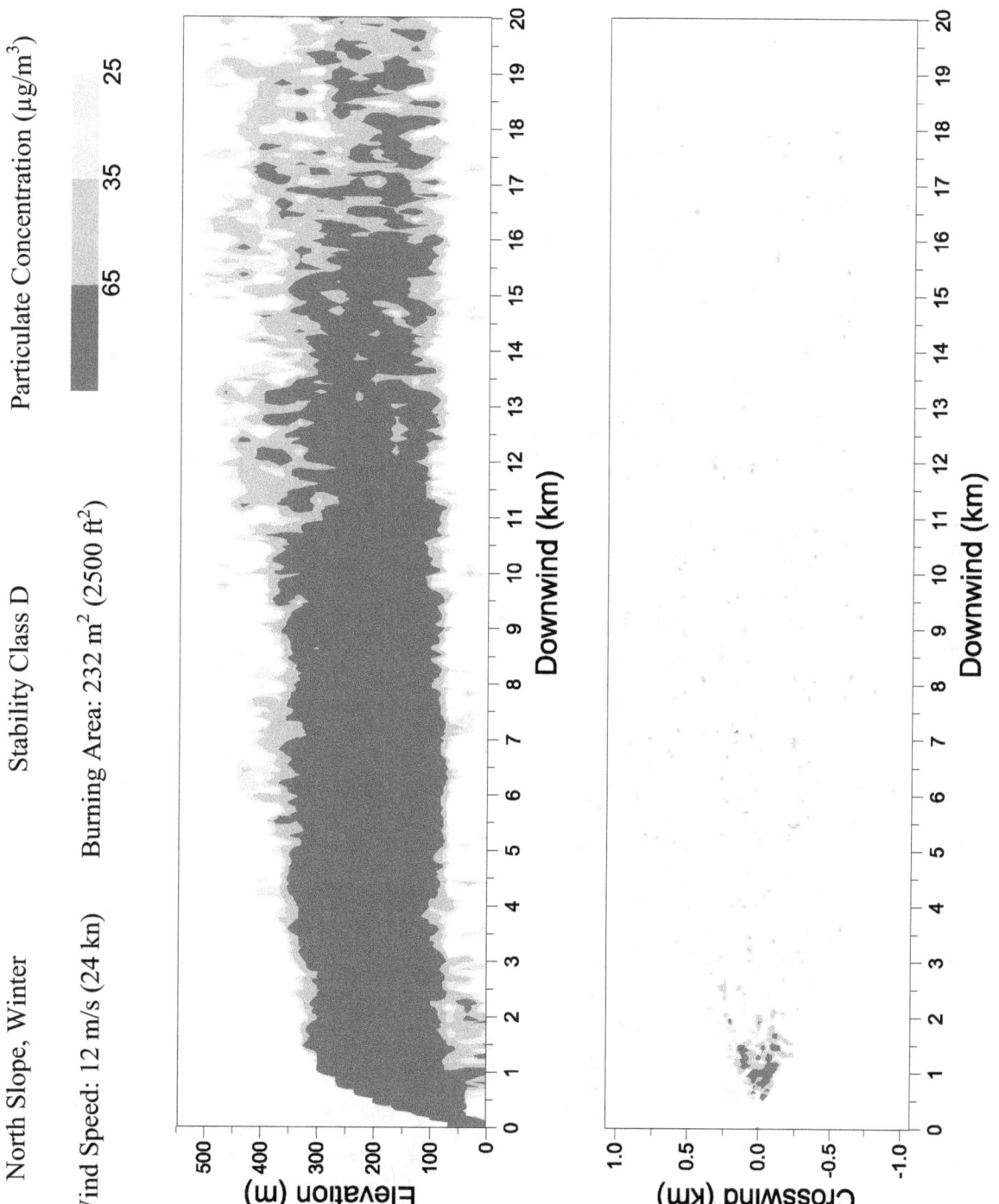

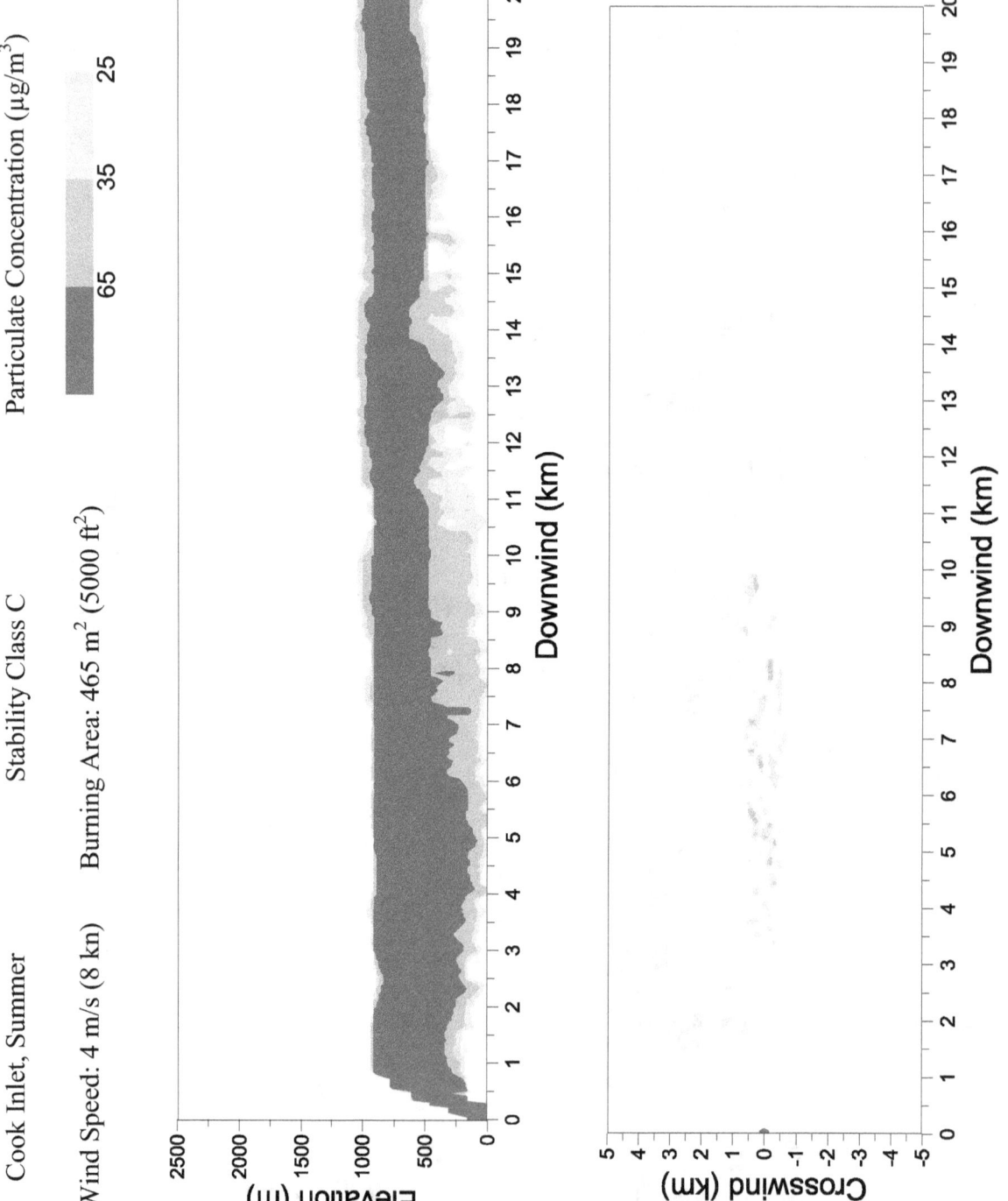

Cook Inlet, Summer

Stability Class C

Particulate Concentration (µg/m³)

Wind Speed: 4 m/s (8 kn)

Burning Area: 465 m² (5000 ft²)

25

35

65

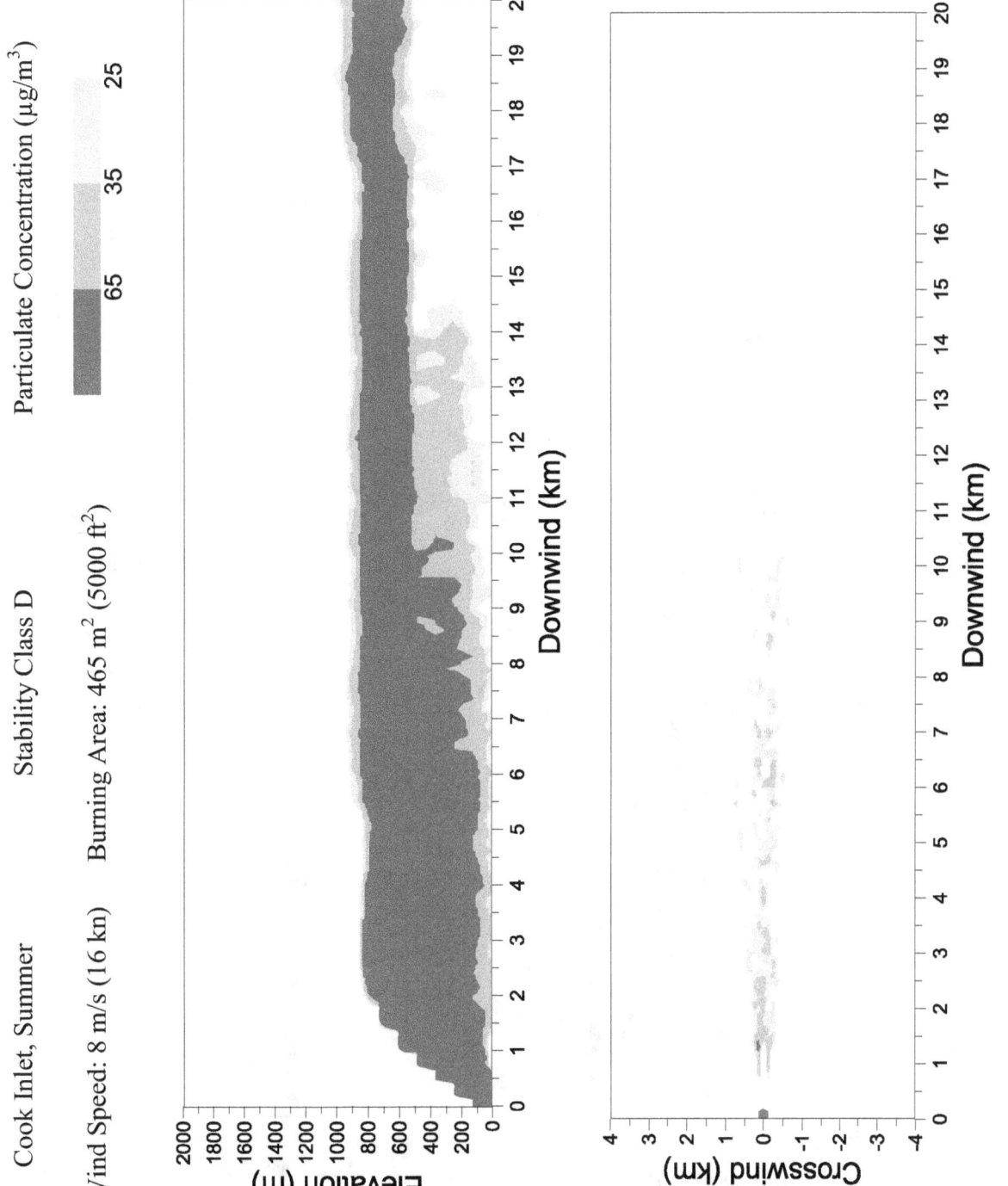

Cook Inlet, Summer

Stability Class D

Particulate Concentration (μg/m³)

Wind Speed: 8 m/s (16 kn)

Burning Area: 465 m² (5000 ft²)

65   35   25

37

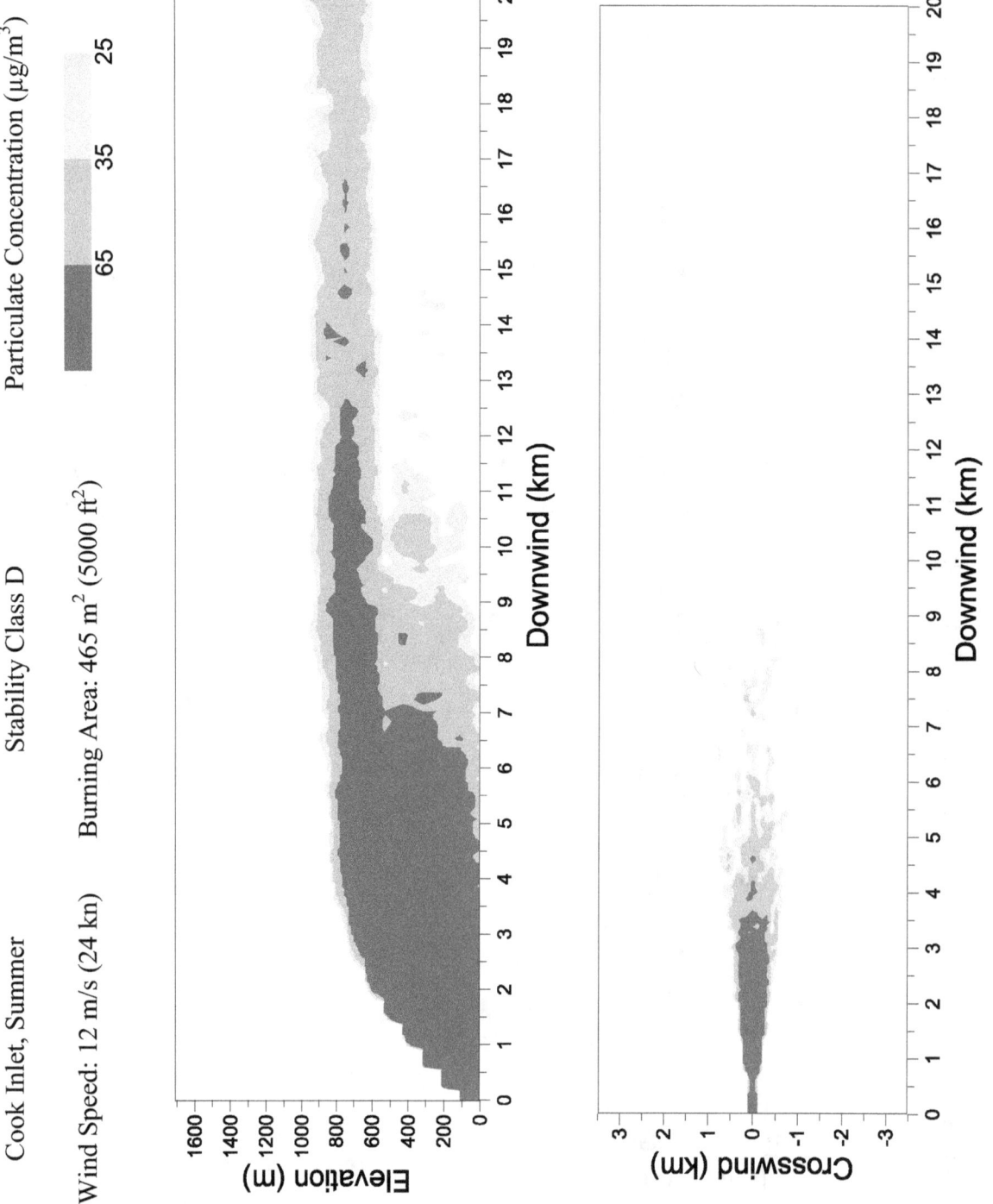

Cook Inlet, Summer

Stability Class D

Wind Speed: 12 m/s (24 kn)

Burning Area: 465 m² (5000 ft²)

Particulate Concentration (μg/m³)

25   35   65

Downwind (km)

Elevation (m)

Crosswind (km)

Downwind (km)

38

Cook Inlet, Winter   Stability Class C   Particulate Concentration ($\mu g/m^3$)

Wind Speed: 4 m/s (8 kn)   Burning Area: 465 m^2 (5000 ft^2)

65	35	25	

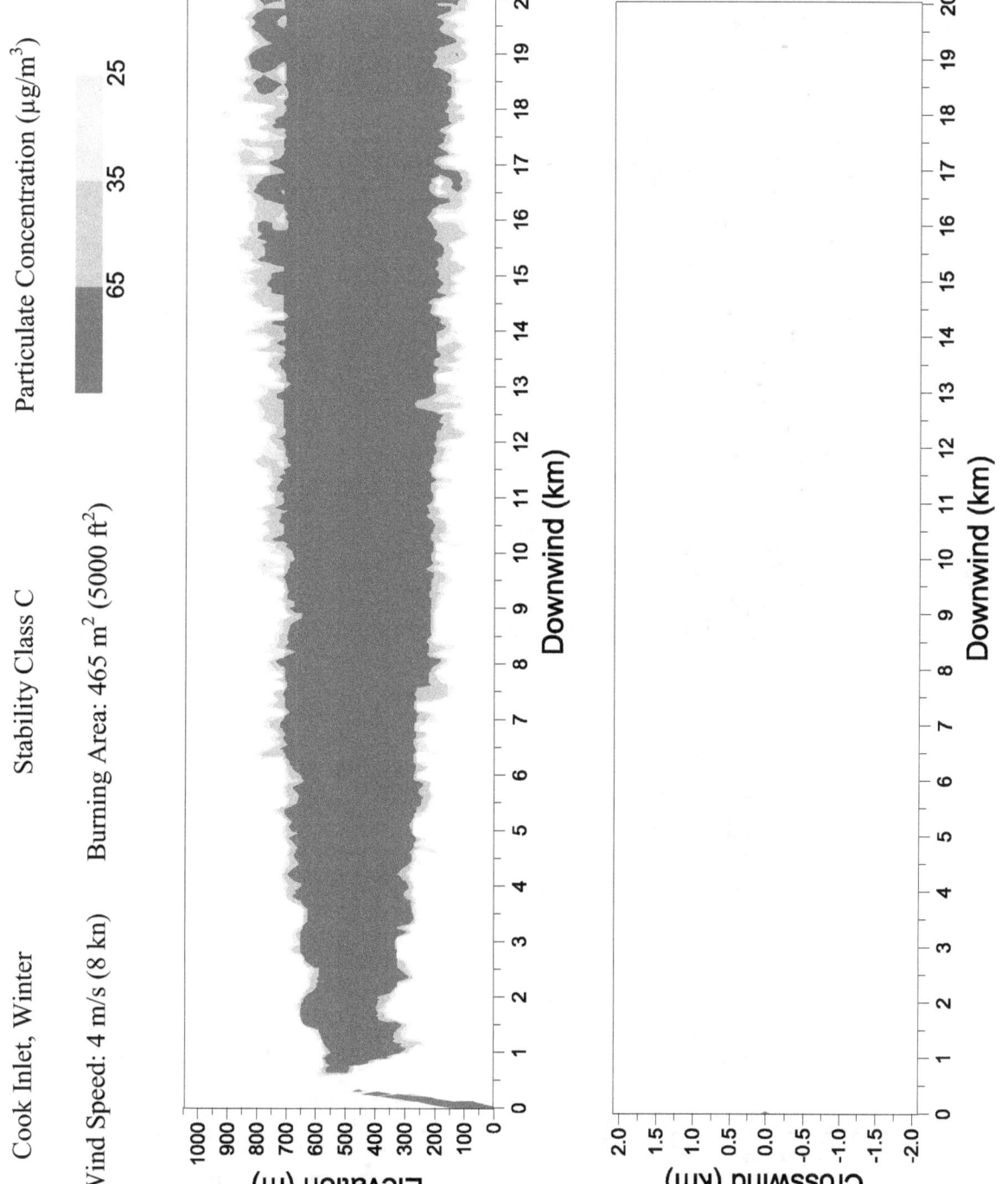

39

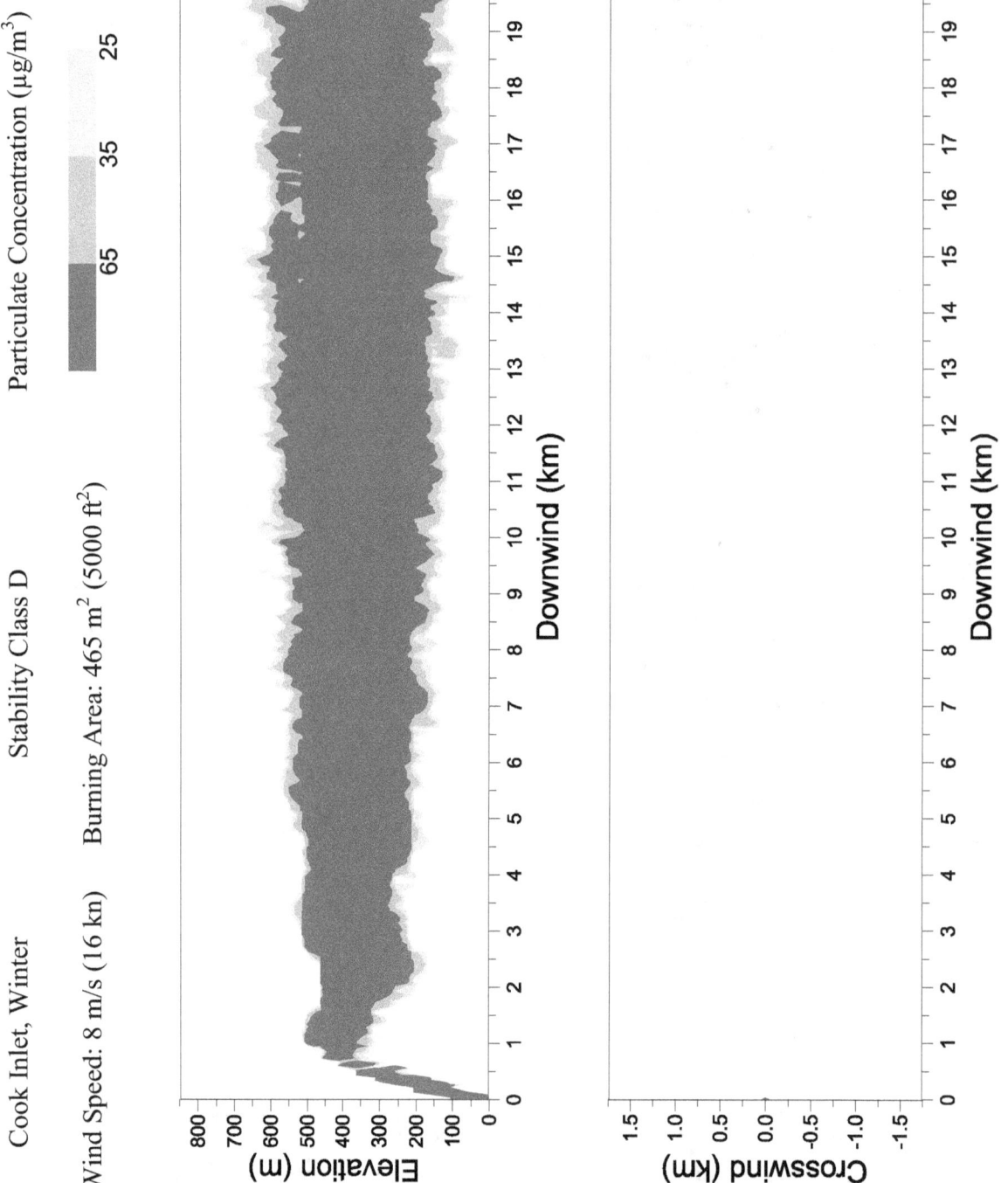

Cook Inlet, Winter   Stability Class D   Particulate Concentration ($\mu g/m^3$)

Wind Speed: 8 m/s (16 kn)   Burning Area: 465 m^2 (5000 ft^2)

25   35   65

Cook Inlet, Winter   Stability Class D   Particulate Concentration (µg/m^3)

Wind Speed: 12 m/s (24 kn)   Burning Area: 465 m^2 (5000 ft^2)

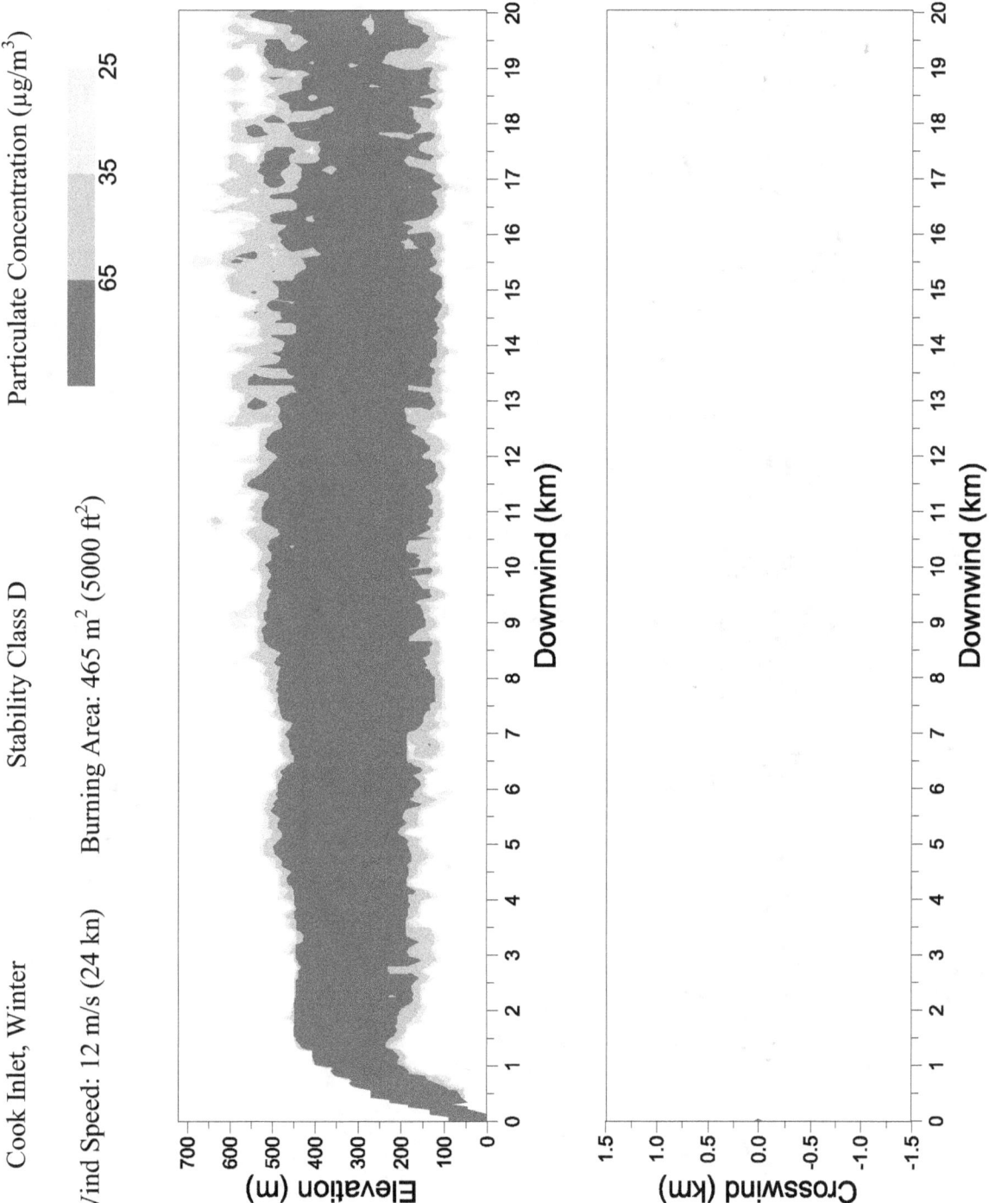

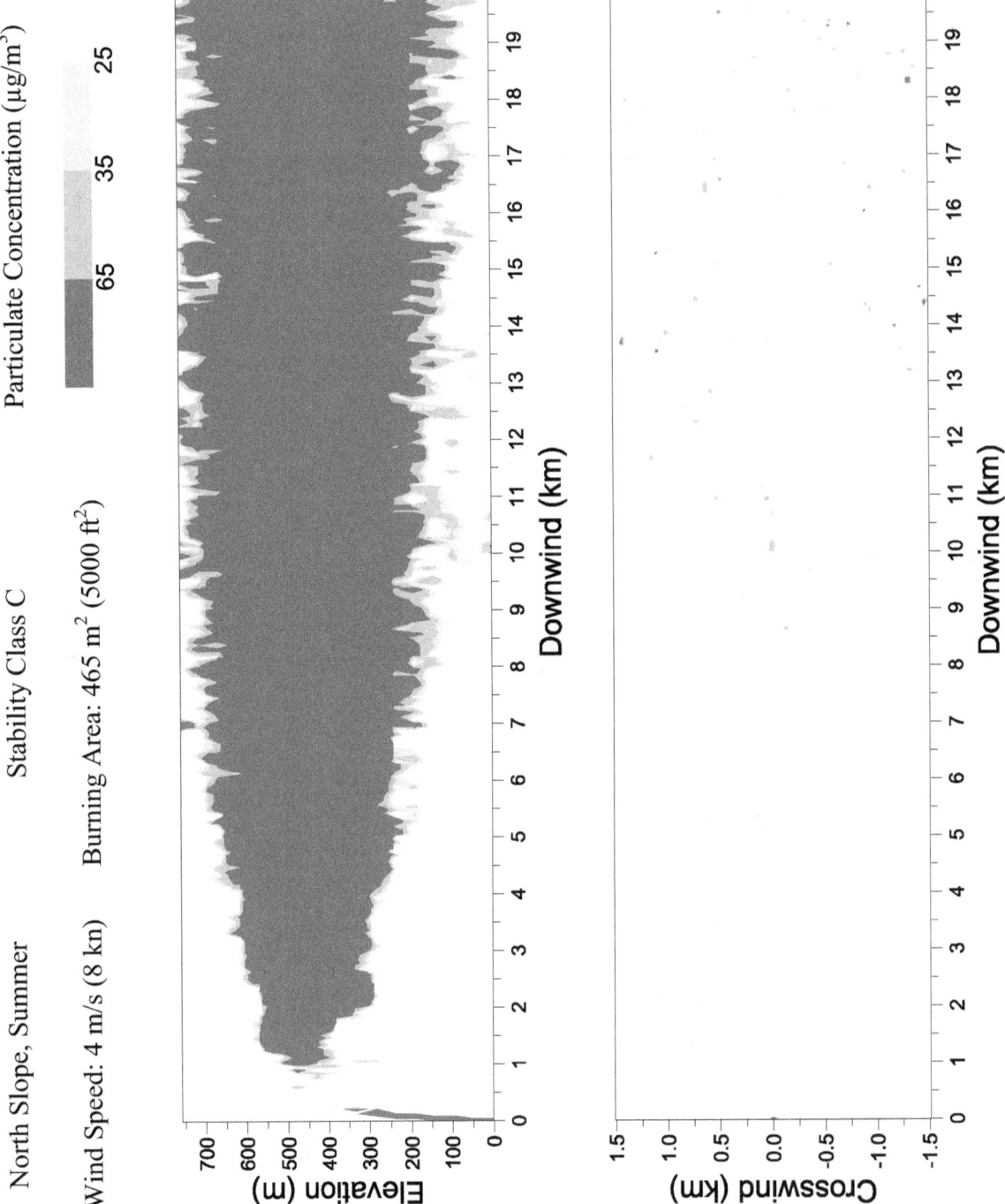

North Slope, Summer   Stability Class C   Particulate Concentration (μg/m³)

Wind Speed: 4 m/s (8 kn)   Burning Area: 465 m² (5000 ft²)

65   35   25

42

North Slope, Summer   Stability Class D   Particulate Concentration (μg/m³)

Wind Speed: 8 m/s (16 kn)   Burning Area: 465 m² (5000 ft²)

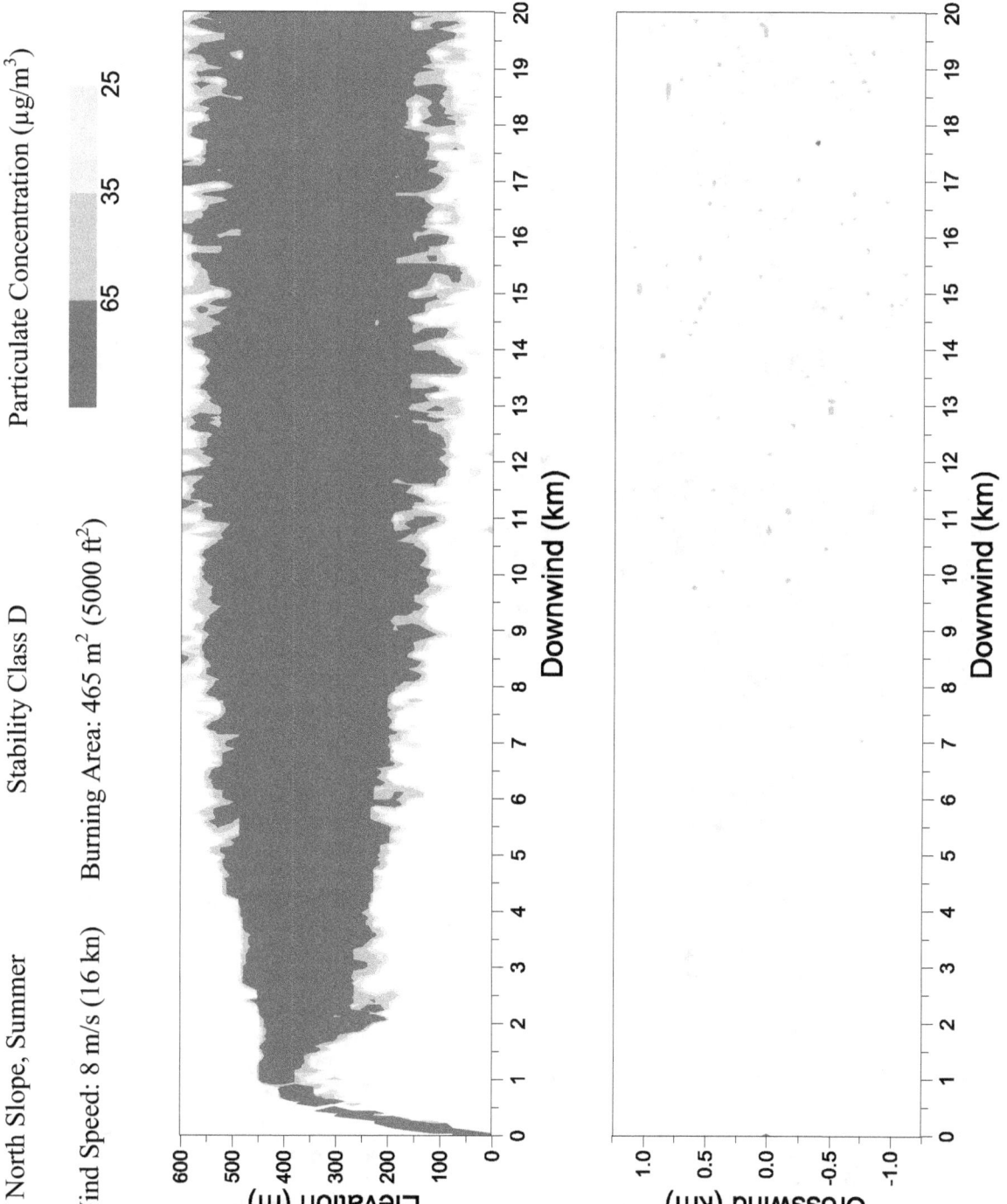

43

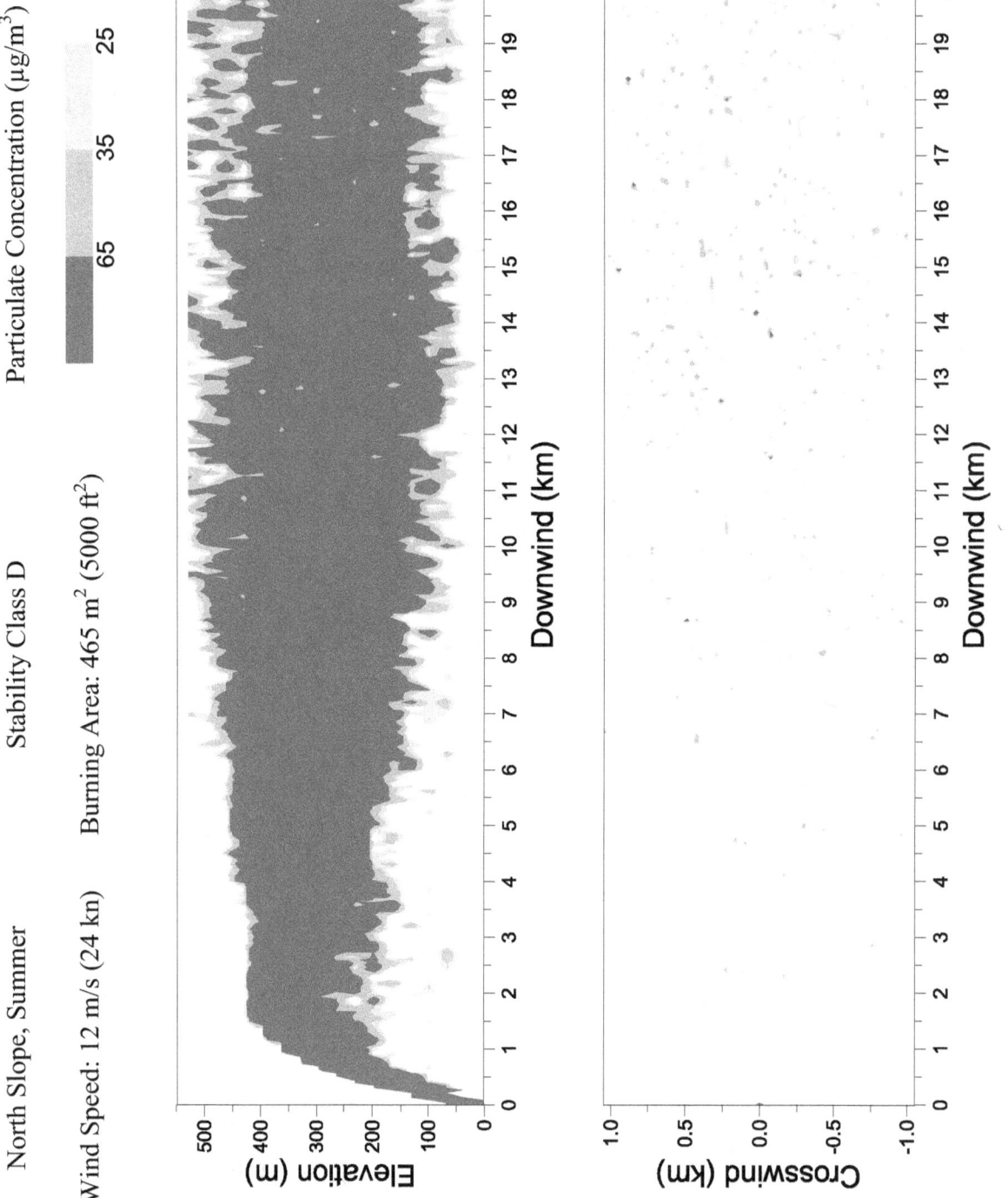

North Slope, Summer   Stability Class D

Wind Speed: 12 m/s (24 kn)   Burning Area: 465 m^2 (5000 ft^2)

Particulate Concentration (μg/m^3)

65   35   25

North Slope, Winter   Stability Class C   Particulate Concentration (μg/m³)

Wind Speed: 4 m/s (8 kn)   Burning Area: 465 m² (5000 ft²)

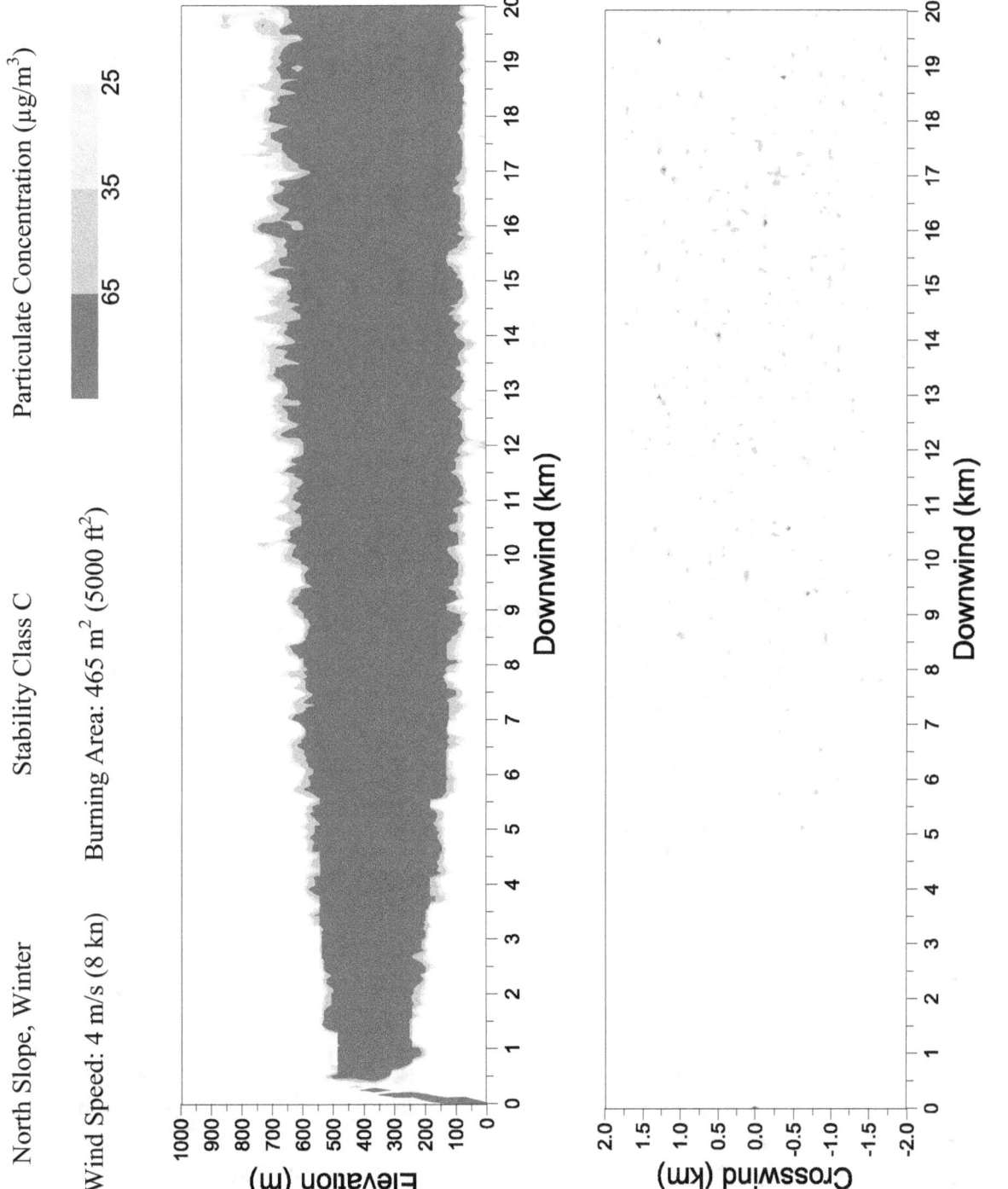

45

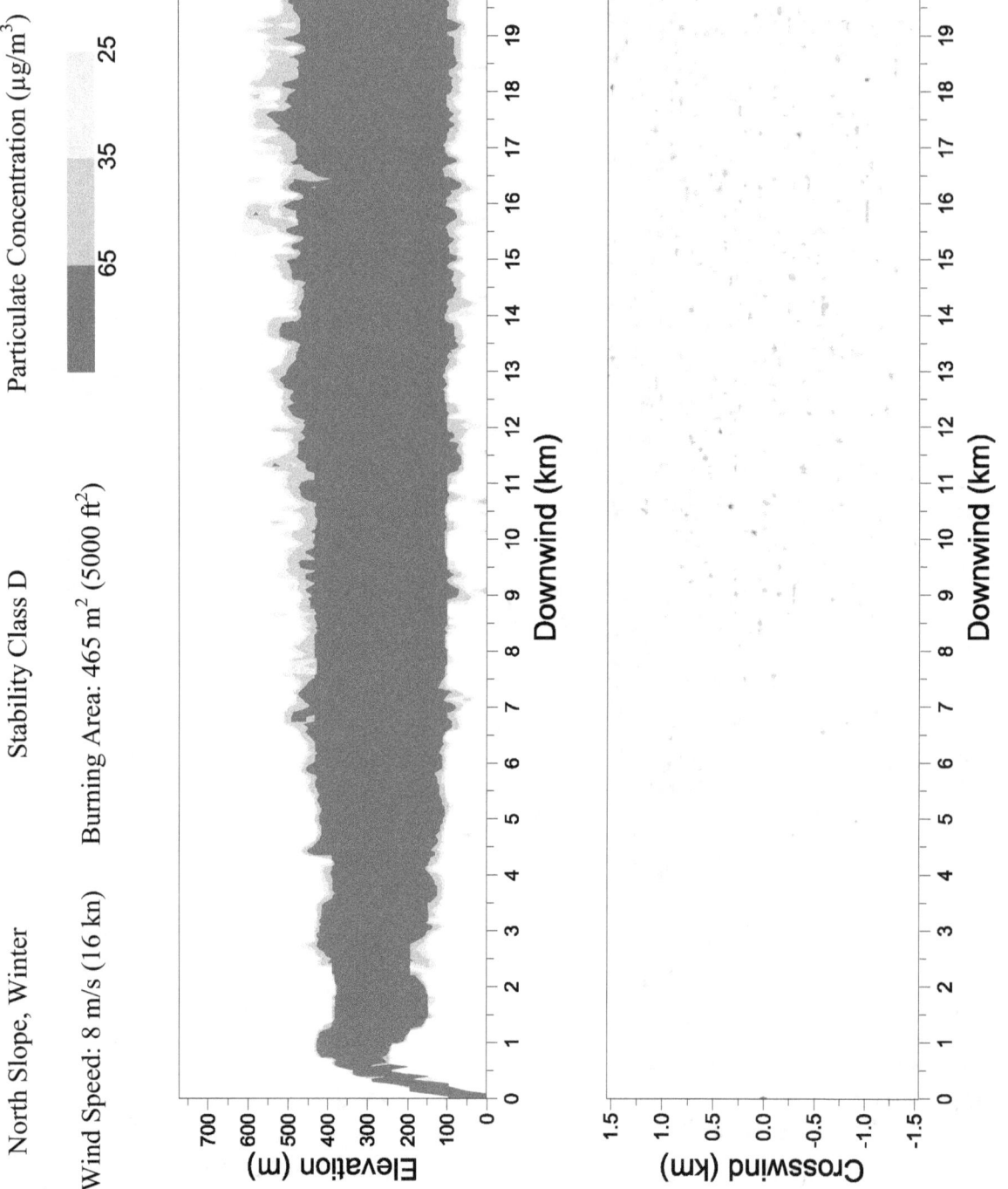

North Slope, Winter   Stability Class D   Particulate Concentration (µg/m³)

Wind Speed: 8 m/s (16 kn)   Burning Area: 465 m² (5000 ft²)

65   35   25

North Slope, Winter   Stability Class D   Particulate Concentration ($\mu g/m^3$)

Wind Speed: 12 m/s (24 kn)   Burning Area: 465 m^2 (5000 ft^2)

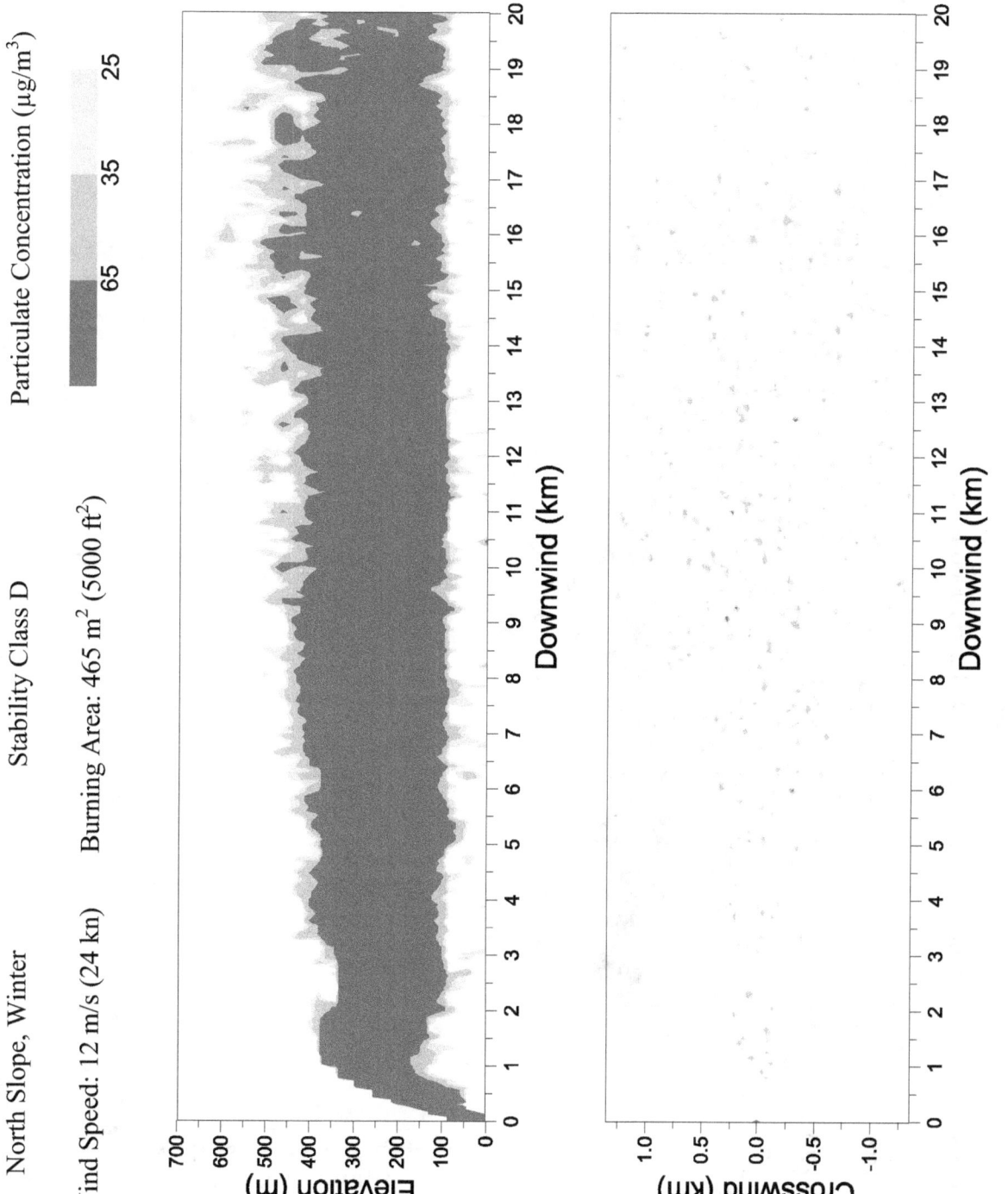

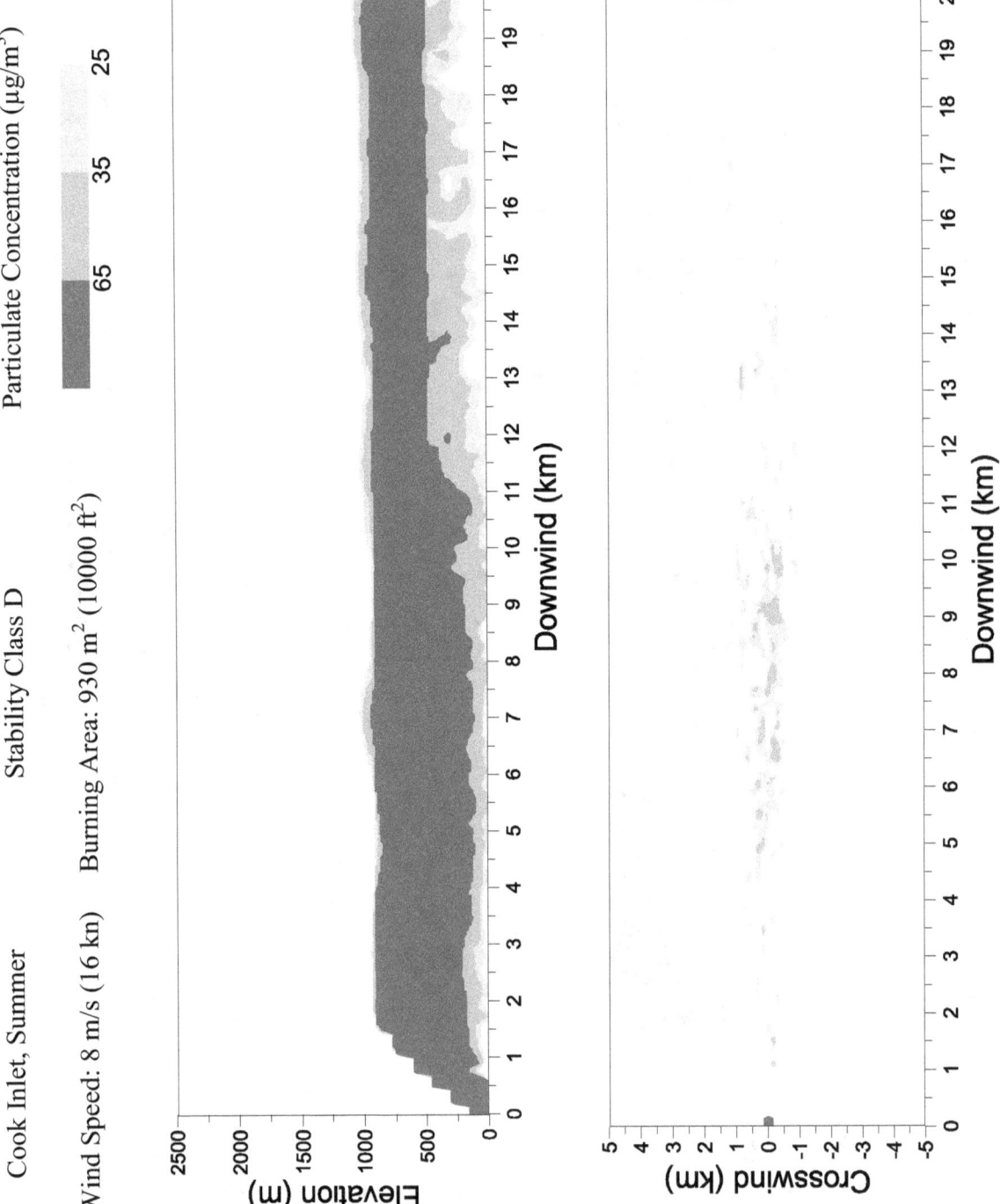

Cook Inlet, Summer   Stability Class D   Particulate Concentration (µg/m³)

Wind Speed: 8 m/s (16 kn)   Burning Area: 930 m² (10000 ft²)

65   35   25

48

Cook Inlet, Winter   Stability Class D   Particulate Concentration ($\mu g/m^3$)

Wind Speed: 8 m/s (16 kn)   Burning Area: 930 m^2 (10000 ft^2)

25   35   65

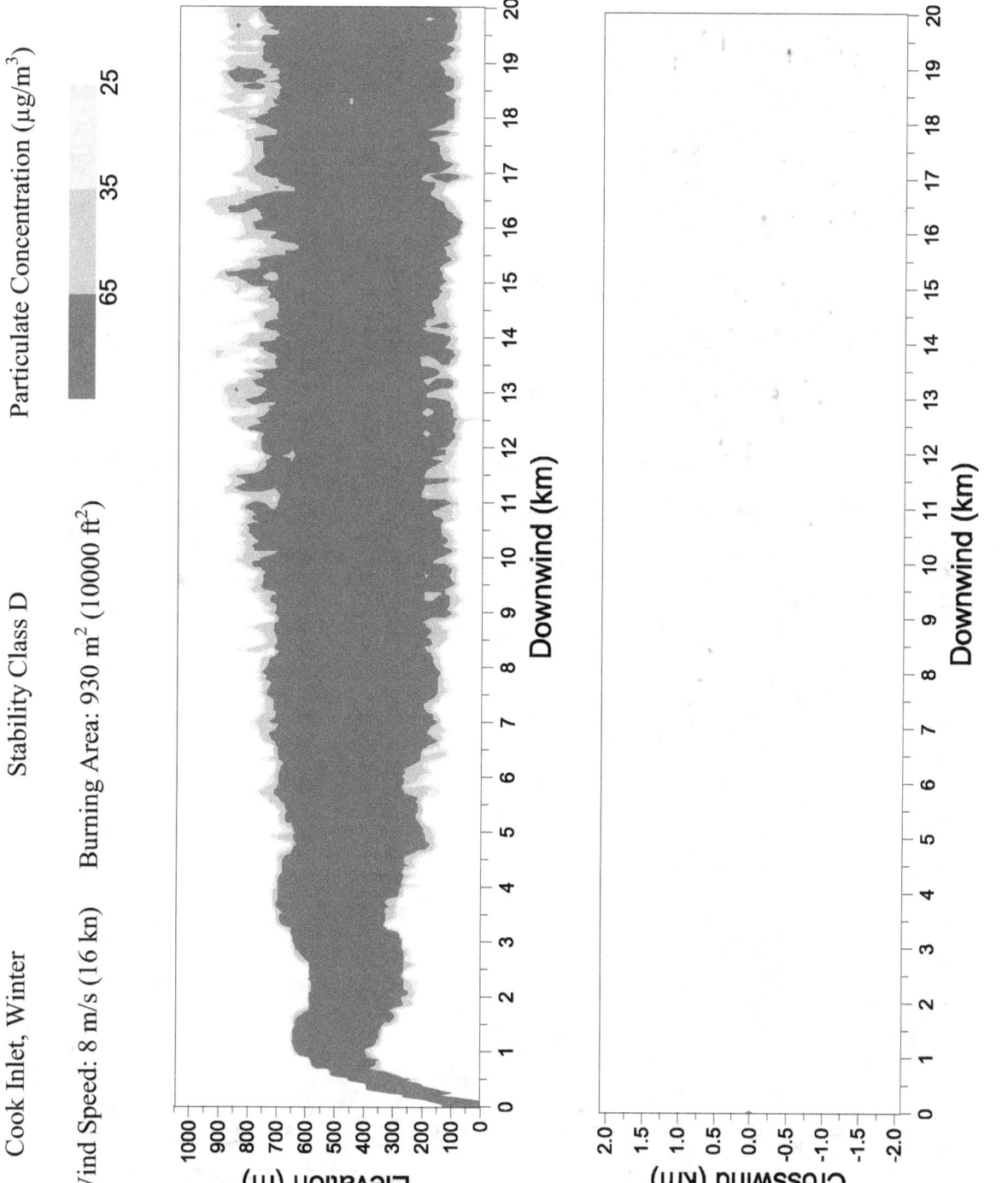

49

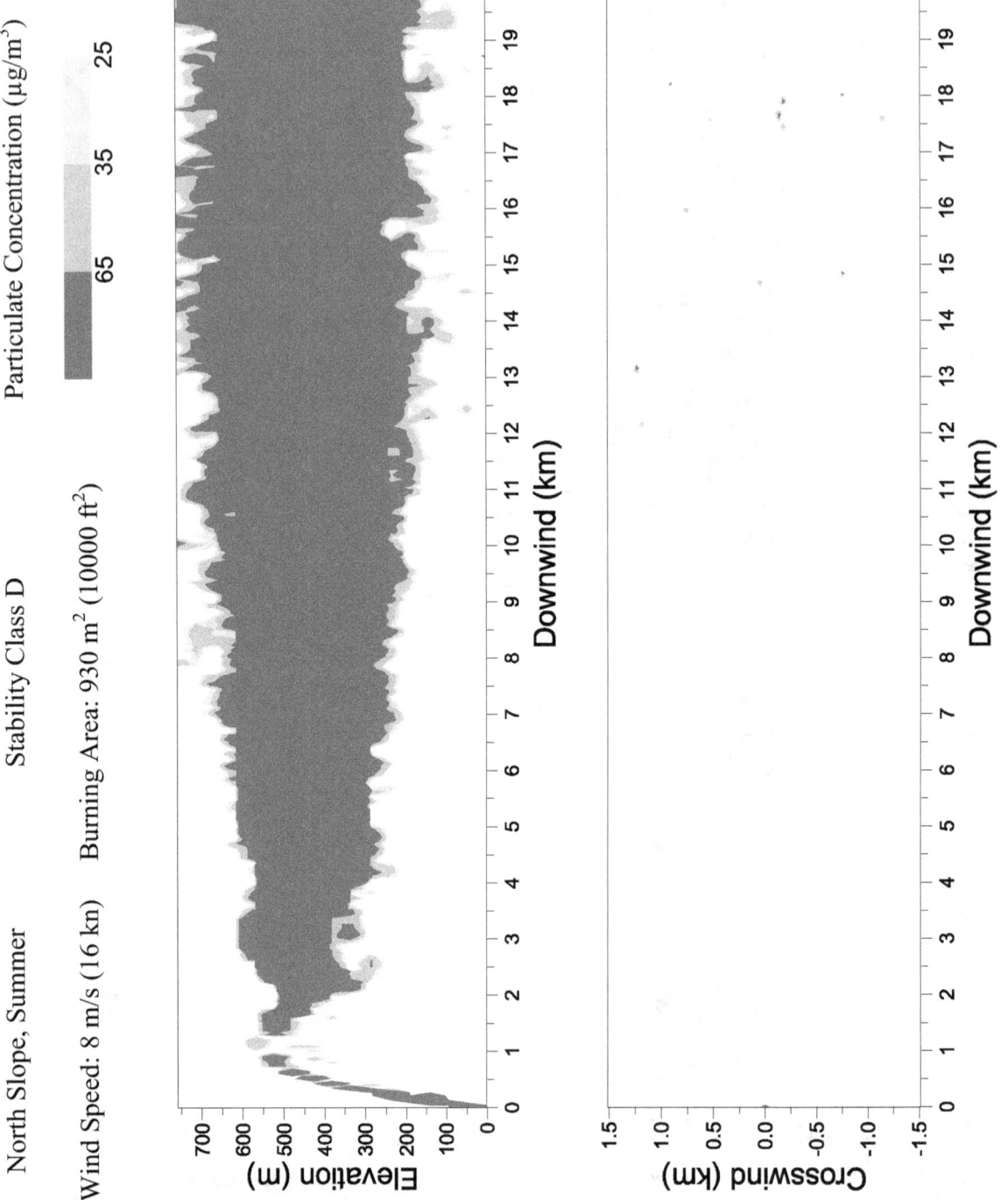

North Slope, Summer   Stability Class D   Particulate Concentration ($\mu g/m^3$)

Wind Speed: 8 m/s (16 kn)   Burning Area: 930 m^2 (10000 ft^2)

25   35   65

North Slope, Winter   Stability Class D   Particulate Concentration (μg/m³)

Wind Speed: 8 m/s (16 kn)   Burning Area: 930 m² (10000 ft²)

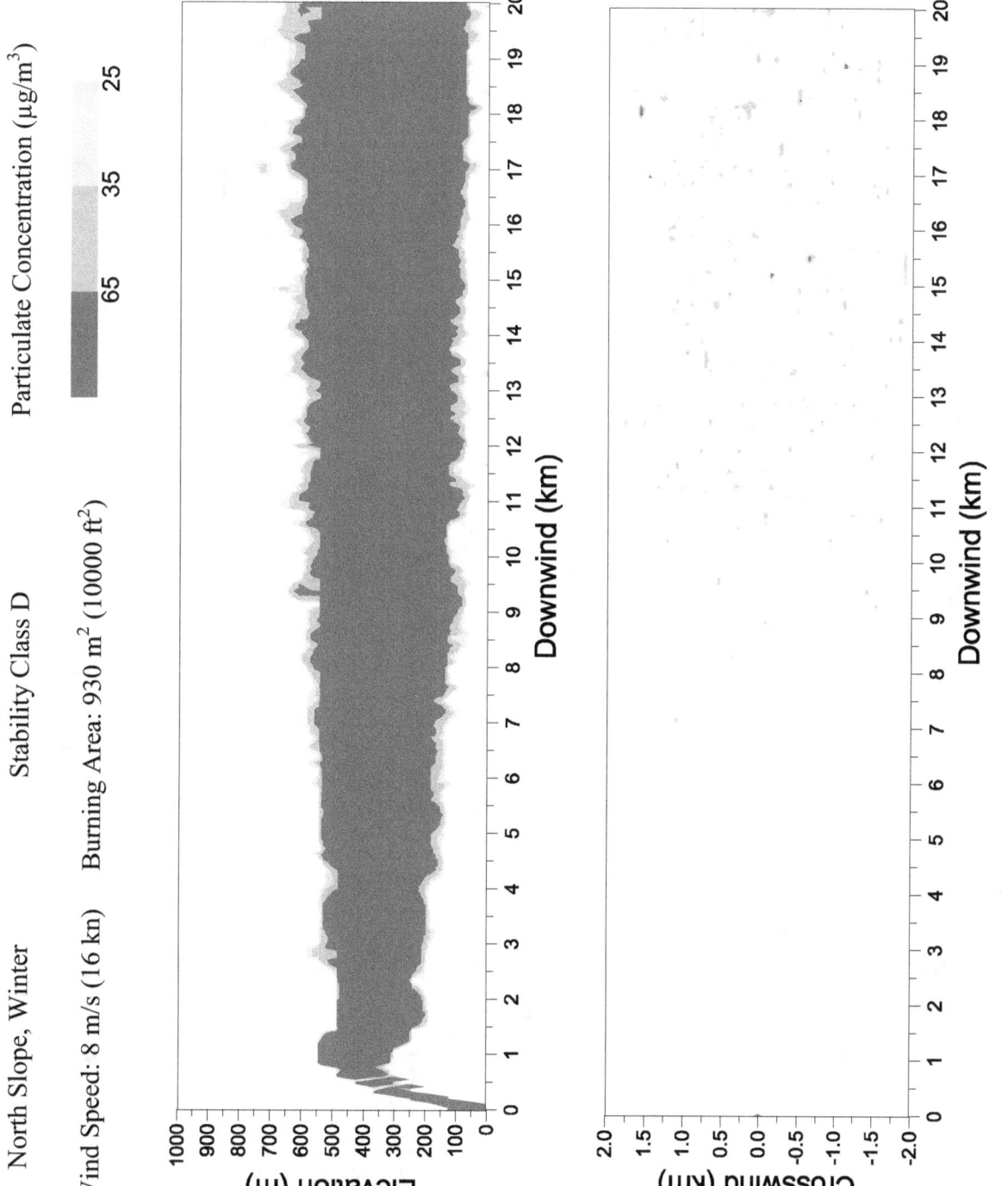

www.ingramcontent.com/pod-product-compliance
Lightning Source LLC
Chambersburg PA
CBHW081852170526
45167CB00007B/2987